# Making an Eight day Longcase Clock

# Alan Timmins M.B.H.I.

**TEE Publishing**
Leicestershire, England

ISBN 0 905100 37 9

© Alan Timmins 1981
© TEE Publishing 1981

All rights reserved. No part of this publication may be reproduced, stored in a retrieval system, or transmitted, in any form or by any means, electronic, mechanical, photocopying, recording or otherwise, without the prior permission of TEE Publishing.

Set in 9 on 9½ point Times by Print Design & Reproduction Ltd, Leicester, and printed in Great Britain by Taylor Bloxham Ltd, Leicester for TEE Publishing, 216 Coventry Road, Hinckley, Leicestershire, England.

# Contents

Foreword .................................................................... 5
Preface ...................................................................... 7
Chapter 1: Description of movement. Material cutting list ...................... 9
Chapter 2: Notes on tools and equipment. Plates and pillars ................... 14
Chapter 3: Barrels and arbors, ratchets, springs and slip washers ............. 18
Chapter 4: Cutting ratchet teeth, great wheels, train wheels and collets ...... 22
Chapter 5: Arbors and pinions ................................................. 27
Chapter 6: Depthing and planting the trains ................................... 31
Chapter 7: The escapement, pendulum, weights and pulleys, testing ............. 35
Chapter 8: The motion work .................................................... 44
Chapter 9: Rack and snail, rack spring, fly ................................... 48
Chapter 10: Bell hammer and spring, lever work, gathering pallet .............. 53
Chapter 11: Planting levers on front plate, setting up strike work ............ 57
Chapter 12: The dial and chapter ring ......................................... 61
Chapter 13: Seconds ring, dial feet, hands .................................... 69
Chapter 14: Datework .......................................................... 73
Chapter 15: Finishing and cleaning, final assembly ............................ 78
Chapter 16: The case .......................................................... 80
Appendix I: Gearing theory and methods of cutting clock gears ................. 93
Appendix II: List of suppliers ................................................ 99

## ACKNOWLEDGEMENTS

As it is always the case, this book would not have been possible without the advice and assistance of many people, most of whom do many hours of tedious work unnoticed behind the scenes. I would like to express my sincere thanks to the following for all they have done during the past months:

Chris Deith, Managing Director of TEE Publishing, for starting it all off.
Norman Smedley, for hours of proof reading and layout work.
Steve Dance, for his patience with the photography.
Bob Davison, for much help and company in the workshop.
Chronos Designs Ltd, Myford Ltd, EME Ltd, Cowells Ltd, for their advice and assistance with machinery.
Goodacre Engraving Ltd, for their beautiful dials.
Richard Shestopal, for lots of bits and pieces and machines.
Derek Knight, for his cases.
Mark Makin Design Associates, for the Cover Design
Finally, to my wife and son, for when I wasn't there!

*To the memory of my father*
*JOHN TIMMINS*
*1913 – 1978*

# *Foreword*

by

## Eliot Isaacs, DLC, TEng(CEI), MITE, MBHI

*Managing Director Chronos Designs Ltd*

To the readers of the magazine *Engineering in Miniature,* the work of Alan Timmins needs no introduction. Indeed, this book, which was first published in serial form in that magazine, has been produced in direct response to the many readers who expressed the wish to have the articles available in more permanent form.

Alan Timmins is a dedicated clockmaker with an enthusiasm for his craft which cannot fail to be conveyed to his readers; he is not only a first rate craftsman, he also has the rare gift of being able to communicate his skill in a manner which even the humblest of beginners cannot fail to understand and appreciate.

Although this book sets out to describe the manufacture of an eight-day rack striking long case clock, in fact it achieves far more than this. The choice of subject was no idle whim — the popular eight-day long case movement was for years the backbone of the traditional English clockmaking industry, and anyone who successfully makes one of these clocks is more than sufficiently equipped to move on to more sophisticated designs. Although the book describes the making of the clock in terms suited to the raw beginner, there is much in these pages that will appeal to all clockmakers, beginner and expert alike. Restorers who need to replace broken or missing parts; repairers who up till now have fought shy of wheelcutting; collectors who have no intention of making clocks but wish to know how the mechanism functions — all will find a wealth of superbly illustrated detail, presented in an eminently readable fashion.

Potterspury
1981

# *Preface*

Over the last few years, there has been a strong revival of interest in Longcase or "Grandfather" clocks, which has caused the inevitable spiralling of prices on the antique market. This in turn has led to many dusty, rusty "family heirlooms" being dug out of cellars and attics all over the country. Many model engineers have been asked to give these fine old clocks a quick overhaul and a good oiling to get them going again, and from this brief foray into the world of clocks, some have developed a much deeper interest in their internal workings.

The late Claude Reeve, who must have been the finest and most prolific "amateur" horologist ever, published several articles on clockmaking and won many exhibition prizes with his beautiful clocks, often of the most complex design and always exquisitely made. Other clockmakers have also published detailed construction articles on Regulator, Skeleton, Bracket, Congreve and Wall clocks, with the occasional electric clock appearing from time to time, but very little, if anything, has been written on the building of a traditional English Striking "Grandfather Clock".

As a result of interest shown by friends, I decided to describe and photograph the making of a typical Longcase movement made in about 1790, which came into my workshop for restoration. I shall describe and clearly illustrate every stage of the project so that anybody with a reasonable knowledge of engineering, and a workshop equipped with the usual range of hand tools, a small lathe, a drilling machine and possibly either horizontal or vertical milling machine, or the accessories to carry out this on the lathe, will be able to complete the project without too much difficulty.

Throughout the book I will describe how to make a few tools and gadgets peculiar to clockmaking which we will need as work progresses, and I have listed suppliers of materials, tools and clock components so that nobody should have any difficulties in obtaining supplies. Wheels and pinions will also be available for those who do not wish to set up to make these or don't feel able to make them, but there really is no mystery involved!

# Chapter 1

*Plate 1. The finished movement and dial.*

## DESCRIPTION OF THE MOVEMENT

Before we get involved in the details of making the clock, I think it important that the mechanism is fully understood, so I will go into some considerable detail in this description. Those of you who have experience of clocks please bear with me whilst I describe the basics to those new to Horology.

Most people who look at an assembled clock movement for the first time tend to think how complicated it all looks, and indeed, it does appear so at first, especially a clock which strikes the hours. However, if the movement is treated as three separate parts, the going or time train, the strike train, both situated between the plates, and the motion work, rack, snail and associated levers which are mounted on the front plate behind the dial, then things begin to look a little more straightforward. I think it easiest therefore, if I draw and describe these three areas individually, and then describe how they relate to each other as the clock runs.

### 1. The Going or Time Train

This consists of four wheels, three pinions, a barrel on which the line is wound, and an "anchor" or recoil escapement which is connected to the pendulum. This clock will have a pendulum which is about 39 inches long and this will swing once a second. This fact gives us the starting point from which we will work out the gear train.

Looking at the drawing (Fig 1), starting from the bottom right is the Great Wheel (D) and Barrel (E) on a common arbor. The Great Wheel is able to revolve on the arbor, and the barrel is fixed to the

## FIG 1.

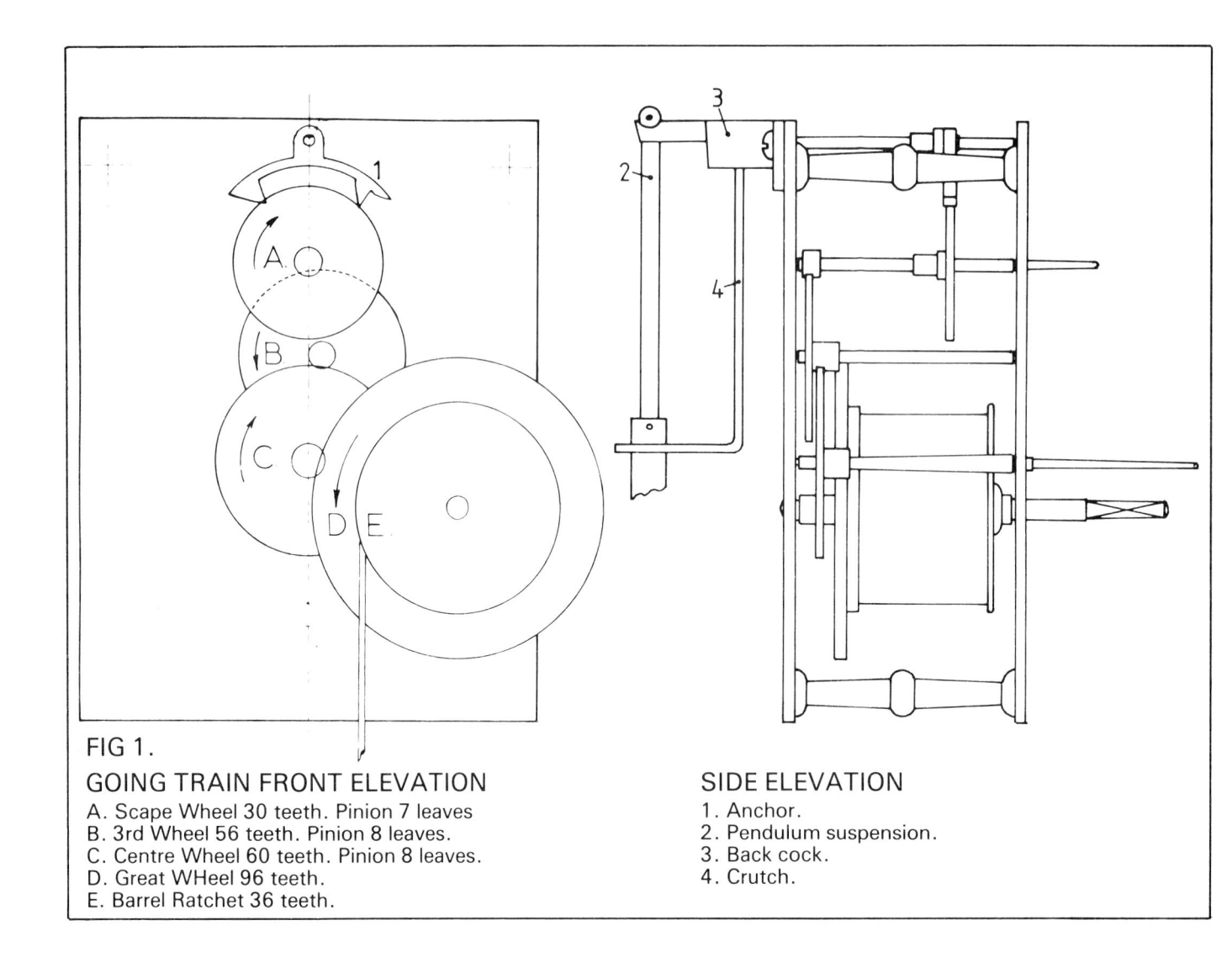

### GOING TRAIN FRONT ELEVATION
A. Scape Wheel 30 teeth. Pinion 7 leaves
B. 3rd Wheel 56 teeth. Pinion 8 leaves.
C. Centre Wheel 60 teeth. Pinion 8 leaves.
D. Great WHeel 96 teeth.
E. Barrel Ratchet 36 teeth.

### SIDE ELEVATION
1. Anchor.
2. Pendulum suspension.
3. Back cock.
4. Crutch.

arbor. The weight is suspended on a line which can be wound round the barrel, and drive is transmitted via a ratchet wheel which is cut on the end flange of the barrel, to a click or pawl fitted to the rim of the great wheel. Thus the line can be wound round the barrel if it is rotated clockwise (viewed from the front), whilst the great wheel remains stationary. When winding is stopped, the click engages a ratchet tooth and the great wheel is driven anti-clockwise by the barrel.

The great wheel drives a pinion on the centre arbor (C). As you will see, the centre arbor also carries a wheel and its front end projects through the front plate. As I will explain later, the minute hand is attached to this projection by a simple friction drive. As the minute hand must revolve once an hour, so must the centre arbor.

Moving now to the top wheel of the going train, we have the escape or 'scape wheel (A). Again this arbor carrying its wheel and pinion projects through the front plate. This projection will carry the seconds hand and must, therefore revolve once a minute. We now have the situation where the centre arbor must revolve once an hour and the 'scape arbor once a minute. Thus, the 'scape wheel revolves 60 times for every rev of the centre arbor, i.e. a gear ration of 60:1. It is possible to provide this reduction in one step but this creates two problems. Firstly, the wheel needed would be too big to fit or would need a lot of very small teeth and secondly, the seconds hand would revolve backwards! The solution therefore is to interpose another wheel and pinion on a common arbor between the centre and 'scape arbors. This extra wheel is called the third wheel (B in Fig 1), and its presence opens up several possible wheel counts to achieve the 60:1 reduction we need. For this clock I have chosen a wheel count which is quite common. This has a centre wheel of 60 teeth, meshing with a 3rd pinion with 8 leaves, giving a ratio of 7.5:1, and a 3rd wheel of 56 teeth meshing with a 'scape pinion with 7 leaves, giving a ratio of 8:1. 7.5 x 8 = 60, which gives us the overall ratio we need.

This leaves us with two more wheels and one pinion to calculate. The 'scape wheel we have already decided must revolve once a minute. This wheel is controlled by the swing of the pendulum which is connected via the crutch to the anchor, as you will see from the drawing. As the pendulum swings, the anchor rocks to and fro, and the two ends or nibs alternately intercept and release teeth on the 'scape wheel, once every second. Each tooth on the 'scape wheel is stopped twice per rev by the anchor so it has to have 30 teeth. To complete the time train, we must decide on the ratio between the centre pinion and the great wheel. It is usual for the great wheel and barrel to revolve once every 12 hours i.e. a 12:1 ratio, and it will be convenient for us to have 96 teeth on the wheel and 8 leaves on the pinion. If the barrel is two inches diameter, about 6" of line will unwind per rev, which equals 8 feet in 8 days, resulting in a very tall clock. If, however, we hang the weight on a pulley and double the line back to the seatboard, we can halve the drop to about 4 feet which is convenient for a longcase clock.

### 2. The Motion Work and Strike Work Mounted on the Front Plate

The Motion work is the simple means by which we can provide our clock with an hour and a minute hand, both rotating in the same direction, geared to each other in the ratio of 12:1. As I said when describing the Going Train, the minute hand is attached to the extended centre arbor via a simple friction drive. In fact, the hand fits over a small square on the end of a tube called the minute pipe. This pipe is an easy sliding fit over the centre arbor extension, has a square spigot on its outer end and is fitted into a wheel with 39 teeth at its inner end. A spring made from a piece of hard sheet brass with a central hole fits over the centre arbor, butting up against the shoulder on the arbor which can be seen in the side elevation Fig 1. The minute pipe is then fitted over the arbor so that the spring is trapped between the shoulder and the wheel on the end of the pipe. This wheel is called the minute wheel. Fig 2.1. The minute hand is held onto the squared end of the pipe by a hand collet or washer, and a pin through the arbor. When the pin is inserted, it pushes the pipe and minute wheel back and slightly compresses the friction spring. This causes the hand to be driven from the arbor via the spring, but enables the hand to be moved independently of the arbor for time setting purposes.

The hour hand is also fitted to a pipe or tube, arranged to revolve concentrically

**FIG 2.**

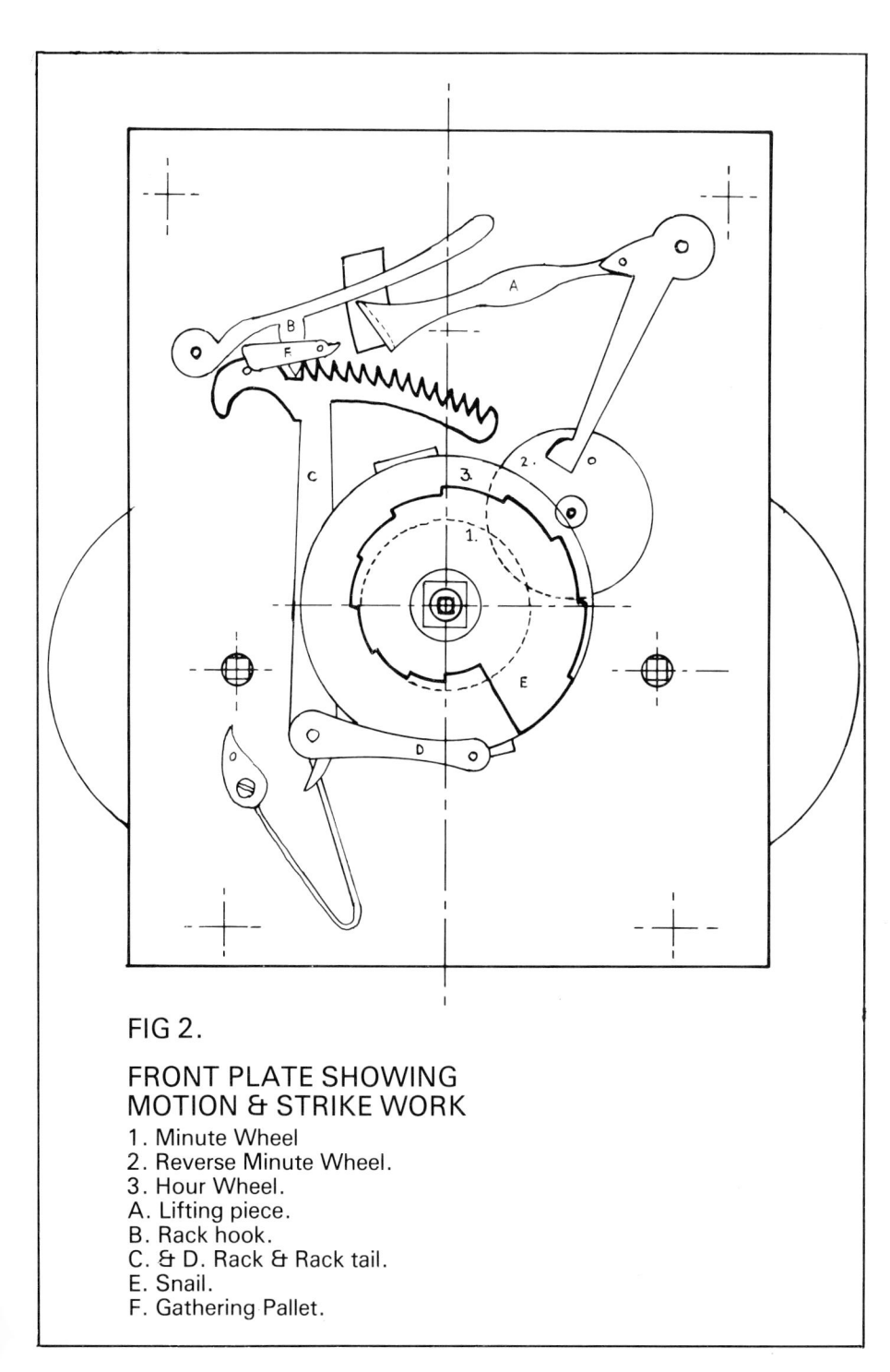

**FRONT PLATE SHOWING MOTION & STRIKE WORK**
1. Minute Wheel
2. Reverse Minute Wheel.
3. Hour Wheel.
A. Lifting piece.
B. Rack hook.
C. & D. Rack & Rack tail.
E. Snail.
F. Gathering Pallet.

with the minute pipe. This is done by fitting a bridge over the minute wheel, screwed to the front plate, carrying a larger diameter tube through which the minute pipe passes. Revolving on this tube is another pipe which carries the hour wheel on its inner end and the hour hand on its outer end. The hour wheel, which for convenience will have 72 teeth, must be geared to the minute wheel in the ratio of 12:1. This is easily done by mounting a wheel and pinion on a stud screwed to the front plate. The wheel, which is called the Reverse minute wheel, meshes with the minute wheel, and has the same number of teeth, i.e. 39. (Fig 2.2). This wheel is mounted on the end of a long pinion which revolves on the stud, meshing with the hour wheel. If the hour wheel is to have 72 teeth, the pinion will have 6 leaves to give us the 12:1 ratio we need. This arrangement also ensures that the two hands revolve in the same direction.

The motion work leads us conveniently into the description of the strike mechanism, as this is triggered by a pin in the reverse minute wheel, and the number of strikes is controlled by a snail mounted on the hour pipe. When the strike train is at rest, it is prevented from running because the tail of the gathering pallet (Fig 2F) rests on a pin which is fixed to the rack (Fig 2C), and is prevented from rotating. The gathering pallet is attached to one of the strike train arbors which projects through the front plate. The first stage of the strike sequence is the warning. This happens a few minutes before the hour. As the reverse minute wheel revolves (once an hour) the pin mounted on it engages the tail on the lifting piece (Fig 2A). The lifting piece itself has an extension which passes through a slot in the front plate, which is lifted into the path of a pin in the rim of the warn wheel (Fig 3B). At the same time, the rack hook (Fig 2B), which rests on the lifting piece, is also lifted, letting the rack move to the left aided by its spring. This releases the gathering pallet which allows the strike train to revolve until the pin in the warn wheel collides with the lifting piece. The number to be struck is determined by the number of teeth on the rack allowed to pass under the gathering pallet. As you will see later, the gathering pallet revolves once per strike, pulling the rack one tooth to the right at each strike until the tail of the pallet is once again arrested by the stop pin on the rack. During striking, the rack is stopped from springing back by the rack hook, which drops into the rack teeth as they are gathered up. The number of teeth allowed to pass the pallet is simply controlled by the rack tail (Fig 2D) which is fixed to and pivots with the tack, and the snail (Fig 2E), which has twelve steps and is fixed to the hour pipe and thus revolves with the hour hand. If the pin in the rack tail falls onto the largest diameter step on the snail, only one tooth on the rack passes under the pallet and the clock will strike one. As the snail revolves, so the rack tail will fall progressively further allowing more teeth past the pallet, until twelve is reached, as in the drawing.

Once the strike train has warned, therefore, the number to be struck is determined and the train is arrested until the pin in the reverse minute wheel passes under the tail of the lifting piece and allows it to drop. This releases the pin on the warn wheel and allows the strike train to run until the gathering pallet is arrested as described earlier.

**3. The Strike Train**
The final section of the clock is the strike train. This is drawn in Fig 3, and is again a common train found in antique clocks. The critical ratios in the train are between arbors B, C & D. As we already know, the gathering pallet which is attached to the end of arbor C must revolve once per strike. Moving downwards, we come to the pin wheel (D) which as its name implies, has a circle of pins projecting from its rim. The bell hammer pivots on an arbor off to the left hand side of the plates with a hammer tail which is in the path of the pins. As the pin wheel revolves, the pins lift and then release the hammer tail thus causing the hammer to strike the bell. In this train, we have 8 pins in the pin wheel so we will get 8 strikes for each rev of the wheel. As arbor C revolves once per strike, the ratio between C and D is 8:1. If we have 56 teeth on the pin wheel, a pinion with 7 leaves on the pallet arbor will give us the correct ratio.

Moving upwards from the pallet wheel we come to the warn wheel. As you will remember, this is the wheel carrying a pin which is arrested by the lifting piece when the clock has warned. This pin must always stop in the same position relative to the lifting piece or striking will be erratic. Thus, the ratio between arbors C and D must be such that the warn wheel revolves an exact number of turns for each rev of the pallet wheel. The actual number is not important, but the normal ratio found is 7:1, which we can achieve by having 7 leaves on the warn pinion and 49 teeth on the pallet wheel. The last arbor at the top of the strike train is the fly arbor. This has a pinion which meshes with the warn wheel and carries the fly

## FIG 3.

### STRIKE TRAIN FRONT ELEVATION

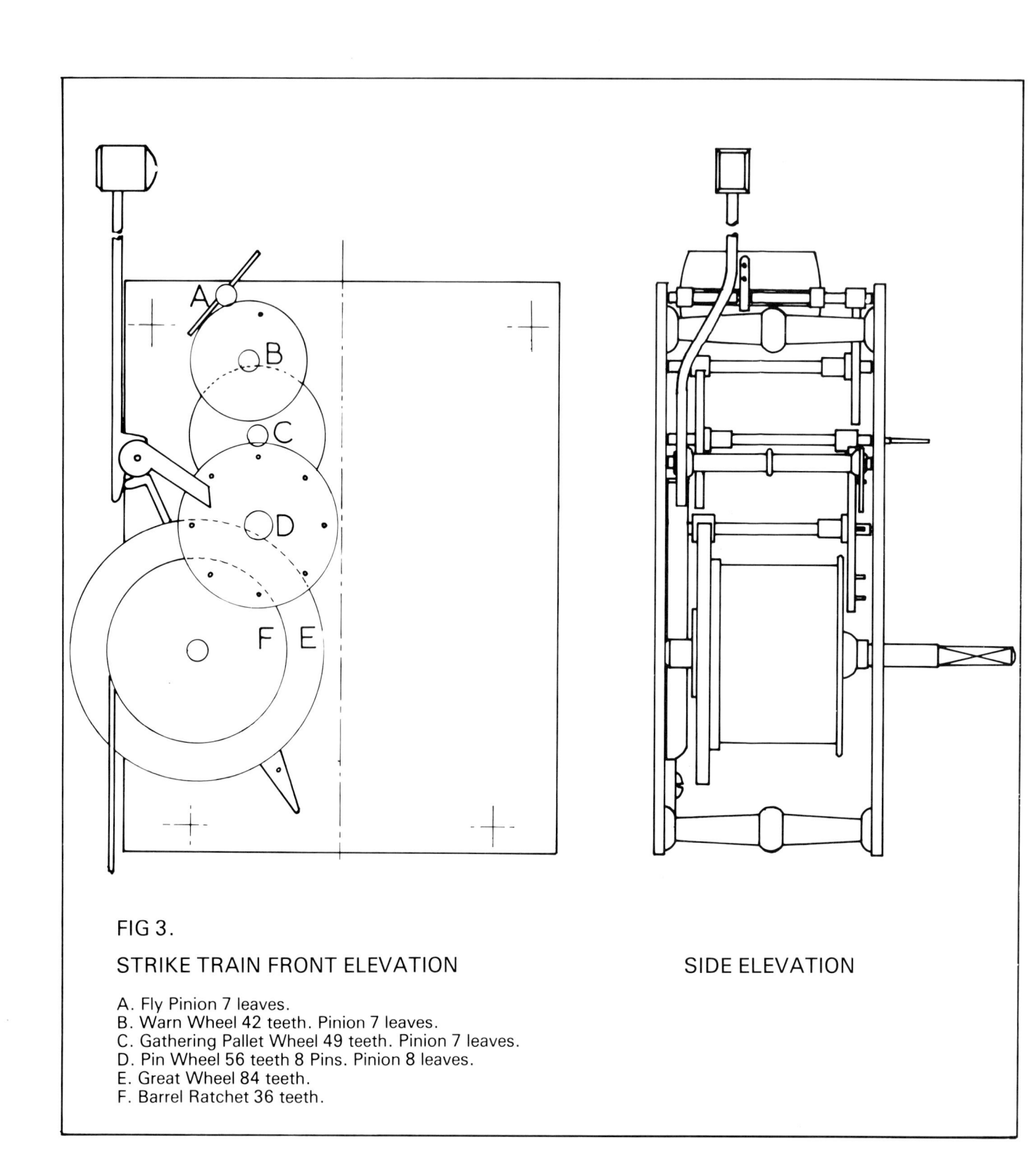

A. Fly Pinion 7 leaves.
B. Warn Wheel 42 teeth. Pinion 7 leaves.
C. Gathering Pallet Wheel 49 teeth. Pinion 7 leaves.
D. Pin Wheel 56 teeth 8 Pins. Pinion 8 leaves.
E. Great Wheel 84 teeth.
F. Barrel Ratchet 36 teeth.

### SIDE ELEVATION

which is a simple two blade air brake. The fly is not fixed to its arbor but is able to revolve on it, held in position by a small leaf spring which engages a groove in the arbor. The spring is stiff enough to drive the fly, but will let it revolve a little when the wheel train is suddenly stopped when striking is finished. The purpose of the fly is, of course, to control the speed at which the clock strikes, so the gear ratio is not of vital importance within limits. Ratios of 6, 7 and 8:1 are common, and as we will be making 7 leaf pinions anyway, we will use 6:1, with 42 teeth on the warn wheel and 7 leaves on the fly pinion. This leaves us with the great wheel and pin wheel pinion. We want the two weights to drop at a similar rate so that we can wind both the strike and time trains at the same time each week. As I said earlier, the pin wheel is to have eight pins and as there are 78 strikes every 12 hours, the pin wheel will revolve 9¾ times in 12 hours or 19½ times per day. For aesthetic reasons, we want the two barrel arbors to be the same height in the plates and equally spaced either side of the centre line of the plates so that we can get a nice symmetrical dial layout. We also want to use the same size wheelcutter as the rest of the clock if possible. With an 8 leaf pinion on the pin wheel arbor, a great wheel of 84 teeth of the same module as the other wheels will put the winding arbor in the place we want it and give us a ratio of 10.5:1, which means that the barrel will revolve near enough twice per day for our purposes.

That completes the description of the movement. I apologise for the length of this description, but as I said earlier a thorough understanding of the "works" is vital, so if you are new to horology, read it a few times and inwardly digest!

## MATERIAL CUTTING LIST

In compiling this list, I have assumed that most people will have odds and ends of material offcuts which they can use to make mandrels etc. Parts such as rack springs, bell stands, hammer springs and crutches are available from several dealers who supply replacement parts for antique clocks. These are supplied in the rough to be finished by the individual to suit the clock. I have included the materials to make such items, but some of you may decide to buy available items.

I also intend to give details of two methods of making the arbors and pinions.

Traditionally, arbors were made from pinion wire which had the leaves pre-formed. Unfortunately, pinion wire is no longer available, so we have to machine the pinions on the arbors. This is a fairly easy operation if you have all the equipment, but it can be tricky for those not so well equipped. When I first started making clocks, I used to machine pinion heads on the end of a piece of rod, which I then drilled through so that I could soft solder them onto silver steel arbors. This means that the arbors can be made from stock size silver steel which is easy to hold in collets to machine the pivots. In this clock, the arbors will be $1/8''$ and $3/32''$ dia. The 8 leaf pinions can safely be drilled and reamed $1/8''$, and the 7 leaf pinions can be drilled just under $3/32''$ for silver steel. This method saves a lot of tricky turning and is invisible if done properly. If you decide to use this method, you need about 4 lengths each of $1/8''$ and $3/32''$ dia silver steel with about a foot of the $5/16''$ dia. steel. If you intend to make the pinions solid on the arbors, I advise you to try to obtain free cutting silver steel, as this machines very easily and is much kinder on carbon steel pinion cutters.

The specification of this steel is KEA 108 and it is available from the suppliers listed.

---

### MATERIAL CUTTING LIST
(N.B. A small allowance has been made for wastage)

| **SHEET BRASS** (Compo or Engraving Brass BS 2870 CZ 120) | | |
|---|---|---|
| PLATES | 2 off | $5 1/8''$ x $6 1/2''$ x $1/8''$ |
| GREAT WHEELS | 1 off | $3 1/8''$ diameter x $3/16''$ |
| | 1 off | $3''$ diameter x $3/16''$ |
| WHEELS | 1 off | $2''$ diameter x $1/16''$ |
| | 1 off | $1 11/16''$ diameter x $1/16''$ |
| | 1 off | $1 7/16''$ diameter x $1/16''$ |
| | 1 off | $1 3/4''$ diameter x $1/16''$ |
| | 2 off | $1 7/8''$ diameter x $1/16''$ |
| | 2 off | $2 7/16''$ diameter x $1/16''$ |
| | 2 off | $1 3/8''$ diameter x $1/16''$ |
| | 2 off | $1 3/8''$ diameter x $3/32''$ |
| BARREL ENDS | 2 off | $2 5/16''$ diameter x $3/16''$ |
| | 2 off | $2 3/16''$ diameter x $3/16''$ |
| MISCELLANEOUS PARTS | 1 off | $6''$ x $6''$ x 22 SWG |
| BRASS STRIP | $1/2''$ x $1/8''$ | $6''$ long |
| BRASS ANGLE | $3/4''$ x $3/4''$ x $1/8''$ | $6''$ long |
| BRASS ROD | $1 1/4''$ diameter | $2''$ long |
| | $1/2''$ diameter | $18''$ long |
| | $3/8''$ diameter | $6''$ long |
| | $1/4''$ diameter | $3''$ long |
| | $1/4''$ square | $3''$ long |
| BRASS TUBE | $2''$ O.D. x $1 3/4''$ bore | $3''$ long |

| | | |
|---|---|---|
| BRASS WIRE | 20 SWG | $6''$ long |
| MILD STEEL STRIP | $1/2''$ x $1/8''$ | $6''$ long |
| | $1/2''$ x $1/4''$ | $3''$ long |
| MILD STEEL PLATE | 1 off | $4''$ x $4''$ x $1/16''$ |
| MILD STEEL ROD | $1/8''$ diameter | $6''$ long |
| | $1/16''$ diameter | $12''$ long |
| | $1/4''$ square | $6''$ long |
| SILVER STEEL | $1/2''$ diameter | 1 length @ $13''$ |
| | $1/4''$ diameter | 1 length @ $13''$ |
| | $1/8''$ diameter | 1 length @ $13''$ |
| FREE CUTTING SILVER STEEL | $5/16''$ diameter | $30''$ long |
| GAUGE PLATE (Carbon Steel Strip) | $1''$ x $1/8''$ | $2''$ long |
| SCREW — Steel Cheesehead | 4BA x $1/2''$ long | 2 off |
| | 2BA x $1/2''$ long | 2 off |
| | 8BA x $1/2''$ long | 3 off |

# Chapter 2

Before I start on detailed construction work, I will go into a little more detail about some of the tools and equipment we shall need or find useful.

Starting with hand tools, most model engineers will have many of the tools we need. The usual range of files, needle files, saws etc., are obvious. A set of BA taps and dies will cover all the threads needed, and most will already own a set of $1/16$ to $1/2$ inch drills and a set of number drills. You will also need a couple of 4" or 3" crossing files of No. 4 or 6 cut for crossing out or spoking wheels, a piercing saw for the same job, with a selection of blades for different thicknesses of metal, and a few five sided cutting broaches for opening out pivot holes etc. I make emery sticks by glueing wet or dry paper to various shapes of bits of wood. 320 and 400 grit will be found useful. I will also describe how to make a few punches and riveting stakes, for fitting wheels to their collets, as we need them.

Besides the usual range of lathe accessories, chucks, centres, etc., you will find that a few collets will more than pay for themselves in terms of accuracy and speed of setting up when turning parts which must be concentric. I make quite a few mandrels form $1/2$" dia. mild steel, with all sorts of holes, spigots and threads on the end. They are very useful for holding wheel blanks whilst they are turned to diameter, and I always keep them. It's surprising how quickly the collection grows, and I can usually pick just the thing I need out of the drawer, put it into the collet, mount a wheel blank by its central hole and guarantee to turn the O.D. concentric to the hole. I certainly couldn't do that with my almost new "super precision" three jaw! Common sizes in this clock are $3/32$", $1/8$", $1/4$", $3/8$" & $1/2$". As an added bonus, I made my dividing head to take all my Myford holding devices so that I can swap workpieces around between the two without problems.

To cut wheels and pinions in the lathe is quite feasible. You will need a vertical slide, onto which you can bolt a milling spindle which you can either make or buy. Nothing fancy is needed and I will include a drawing of one when it is needed. The spindle can be driven by a small motor of about $1/10$th horsepower, mounted on a board which can be clamped to the tailstock end of the lathe bed. Drive to the spindle is easily arranged with plastic belting which stretches and is easily joined by heat.

Dividing and associated equipment must have been the subject of more articles in the model engineering press than any other operation that we have to cope with. The most versatile solution is the dividing head, with either 40:1 or 60:1 reduction and a couple of division plates. These are very expensive to buy and are usually quite large units, or are unsuitable for our purposes. After much thought, I made my own to suit my needs, using a lathe changewheel and an acme thread as the reduction gear, mounted in a fabricated body, with a spindle which accepts all the Myford accessories. I hope to prepare an article on this accessory in the near future. Other methods of solving the problem which are especially suited to gearcutting in lathes,centre around mounting either change wheels or a division plate on the back end of the headstock, and a detent mounted on the back cover of the changewheel case. This direct indexing method is in fact a modification of the traditional clockmakers wheelcutting engine, and if division plates are available to cover the number of teeth required, the direct indexing method is more straightforward and quicker than the geared dividing head. Dividing plates especially suited to the needs of the clockmaker are now on the market again at quite reasonable prices, or can be made by any competent model engineer.

The final requirement for wheel cutting is the cutter. Again, these can be bought from horological suppliers. The commercial items are multi-tooth cutters about $3/4$" diameter, nowadays made to the metric module system, with an epicycloidal tooth form. Unlike involute cutters, only one cutter (for each module) will cut a wheel of any number of teeth. Pinions, however, need different cutters for different numbers of leaves because of the more severe angular variations between the teeth of say a 6 and an 8 leaf pinion. Brass wheels can be cut with simple single point fly cutters; indeed, when making replacement wheels for antique clocks I usually make a flycutter so that the tooth shape of the new wheel matches that of the old wheels in the clock, which were of course cut before standards for gear teeth were established and thus vary a great deal from clock to clock. I will describe the method of making these cutters when we start wheelcutting. It is easier to cut pinions with commercial cutters, because the steel used for pinions obviously isn't as freecutting as brass and the slender tips of flycutters are not strong enough to stand the strain.

Once the wheels and pinions are cut, the wheels must be mounted on their arbors and the train "planted" between the plates. For this job, the old clockmakers devised a tool which, if well made, makes the marking out of the pivot holes foolproof. The Depthing Tool, in which adjacent wheels and pinions can be mounted between centres and then adjusted for free running before marking out the centres of the pivot holes, is an invaluable piece of equipment. There are ways of doing the job without the tool, and I shall mention these as we go along. Depthing tools are available from suppliers but are rather expensive. However, articles have appeared in the *Model Engineer* magazine in the past, or you can do as I did, and build one using part machined castings available commercially with all the critical machining done.

All other work on clocks can be done with normal engineering tools and methods. As most people will only perhaps want to make one clock, they will not want to start buying specialist tools for the job. There is always a way round any problem as model engineers know only too well.

## PLATES

The plates are made from two pieces of brass, $1/8$" thick, 5" x $6 3/8$". All sheet brass used in the clock is engraving brass, ordered from suppliers as Compo Brass. This material cuts and drills very well leaving very few burrs, and can be obtained cut to size.

For the plates, obtain two pieces $5 1/8$" x $6 1/2$", as the sheet is guillotine cut and this leaves a small bevel along the edges which must be removed, so we need to allow $1/16$" on each edge to clean up. When you have got your plates, the first stage is to check for flatness. Compo brass is usually pretty good in this respect, and a small amount of distortion is acceptable, as long as the convex surfaces face inwards on the finished clock. So, having checked your brass and clamped the two pieces together with toolmakers clamps convex sides inwards), mark a very light centre line down the long dimension (see Fig. 4). $1/8$" in from the top and bottom edges of the plates, centre punch and drill $1/16$" or No.52, for register pins. Separate the plates and lightly countersink to remove any burrs. Tap a tapered clock pin (or a piece of steel rod filed to a gentle taper in the lathe) into each hole in the front plate, and lightly open out the holes in the back plate with a cutting broach until the two plates will just press together by hand with no shake in the holes. We can now separate the plates at will and put them together again in exactly the same position without problems.

Hold both plates together in the vice (use fibre jaws to prevent marking), and file one long edge straight, removing the guillotine marks. If you have a milling machine you can save time here by milling all four edges square. When you have prepared the first edge, mark out the other three using it as the datum. The dimensions are not critical as long as all is square. File or mill the plates to size then, using a no. 4 or 6 cut hand file, drawfile and finish with a fine emery stick. A sign of good clockmaking is to see nice crisp sharp edges and good flat surfaces, so always pay particular attention to finish.

When the plates are to size, mark out the position of the pillars from the drawing. As you will see, you can have the usual four pillars or an optional fifth which is positioned low down in the centre of the plates. This fifth pillar is found on what are now regarded as the better quality "London Movements" which are much sought after in the antique trade. It's presence is not vital but it does tend to stiffen up the plates in an area of maximum stress, and it is useful when assembling the finished clock as one pin through the central pillar will hold the plates together whilst the positions of wheels in the strike train are checked.

With all pillar holes marked out and centre punched, drill pilot holes through both plates. Although compo brass drills better than other types of brass it still has its problems, and I usually back off the edge of my drills with a few strokes with a slip stone before drilling to prevent any tendency to grab. Remove any burrs which appear

**FIG 4.**
**FRONT & BACK PLATES**

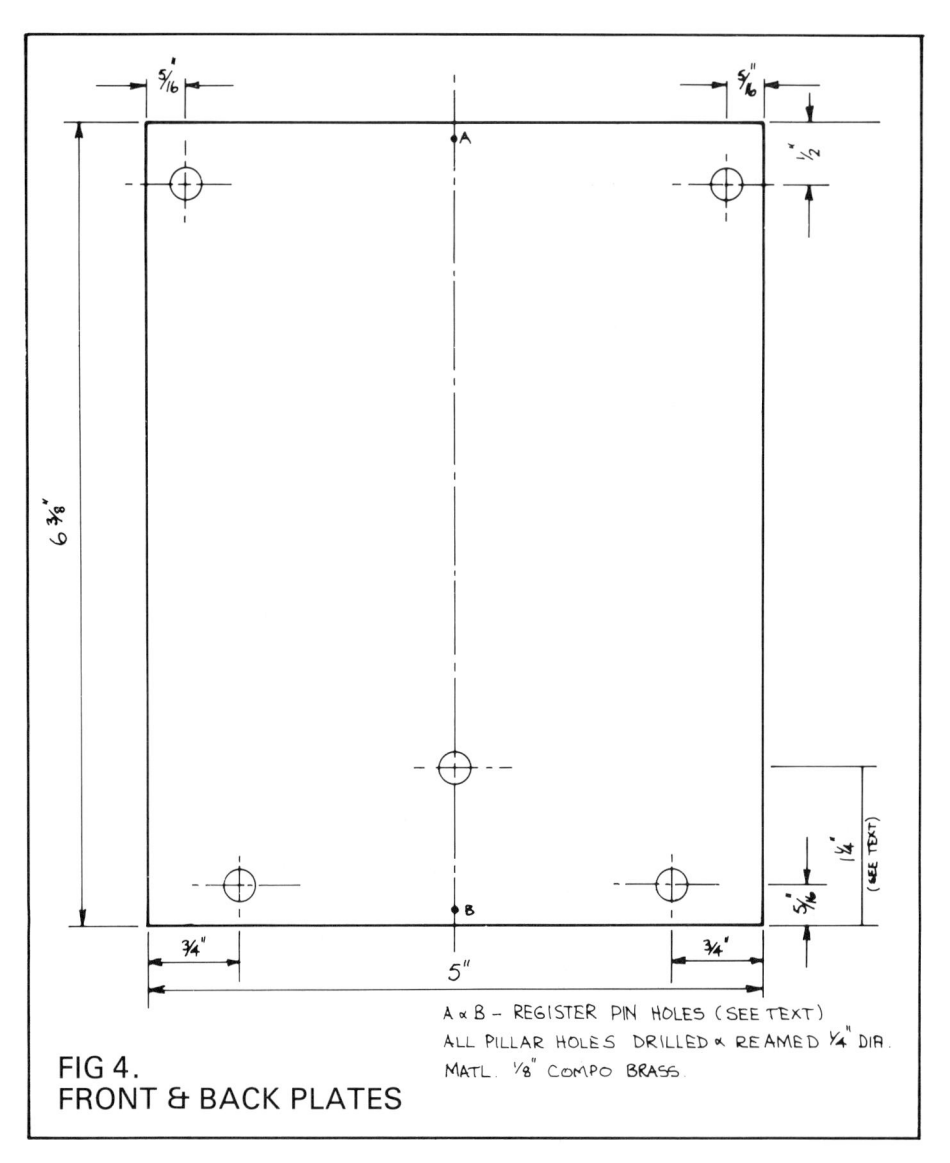

A & B — REGISTER PIN HOLES (SEE TEXT)
ALL PILLAR HOLES DRILLED & REAMED ¼" DIA.
MATL. ⅛" COMPO BRASS.

between the plates as you go along. Open out the holes to ¼" preferably finishing with a reamer.

### PILLARS

For the pillars we need four (or five) lengths of ½" dia. brass each 3" long. Face off both ends in the lathe and centre drill both ends with a BS1 drill (the smallest size). We now come to the first lathe job where a collet makes life a lot easier. the ¼" dia spigots on the ends of the pillars (see Fig. 5) must be concentric. If you haven't got collets, they must be turned between centres, but the decorative turning is not so important, so do the decorative bit first, followed by the long spigot then the short one. If you have a ½" and ¼" collet, proceed as follows:—

After centre drilling both ends with the job held in the ½" collet, turn a spigot on one end ¼" dia. x 3/32" long. It is best to make a small gauge from a piece of ⅛" brass plate with a ¼" hole in it, made with the same drill or reamer you used for the holes in the plates, in case your drill is cutting over size.

The spigots should be a good but not tight fit in the holes, and the shoulders should be slightly undercut so that the full ½" dia. of the pillars will bear on the plates. At this stage I should mention that the traditional method of fixing the pillars to the back plate was to rivet them permanently in place. If you wish to do this, make the short spigots slightly over ⅛" long to allow for riveting. However, when cleaning old clocks I find that riveted pillars make it very difficult to clean the inside of the backplate and the pillars, especially if they need a high polish. On this clock therefore, I intend to fix the pillars to the backplate with screws. Another alternative open to you if you have the taps and dies is to tap the plates with a fine thread (ME, or 40 T.P.I.), and thread the spigots accordingly. In this way, you achieve the appearance of the riveted method but you can unscrew the pillars for cleaning.

The method I use is to drill the short spigot No. 34 and tap 4 B.A. x ½" deep, so I drill each after turning each spigot (you will save a lot of time by doing each operation on all pillars in turn). When you have finished all pillars to this stage, mark off the 2⅜" shoulder length from the shoulder already formed. Replace each pillar in the ½" collet and turn the long spigot again ¼" dia., with the shoulder undercut. At this stage the spigot will be over long to allow for removal of the centre drilling later. The shoulder length is not critical to a few thous. as long as they are all identical. I ensure this by measuring with a Vernier to find the shortest, holding its long spigot in the 3 jaw chuck with the shoulder hard against the jaws and supporting the free end with a revolving centre. Then take a light facing cut across the shoulder at the tailstock end and lock the saddle to the bed. If you now repeat the process on all pillars without disturbing the saddle or top slide then all will be identical. You may need to undercut the shoulders of one or two again, but be careful not to cut to too small a diameter!

Those with the tools and the experience may like to do the decoration by hand turning as the old clockmakers did, but I never seem to be able to get two the same! I ground up two form tools from 5/16" square H.S.S. (see Fig. 6), or you can shape them to suit yourself. The right hand cutting tool (1) forms the radii at the ends up to the shoulders and the left handed tool (2) forms half of the central bulge. Using odd-leg callipers with the pillar spinning in the lathe, mark off ⅛" in from each shoulder and

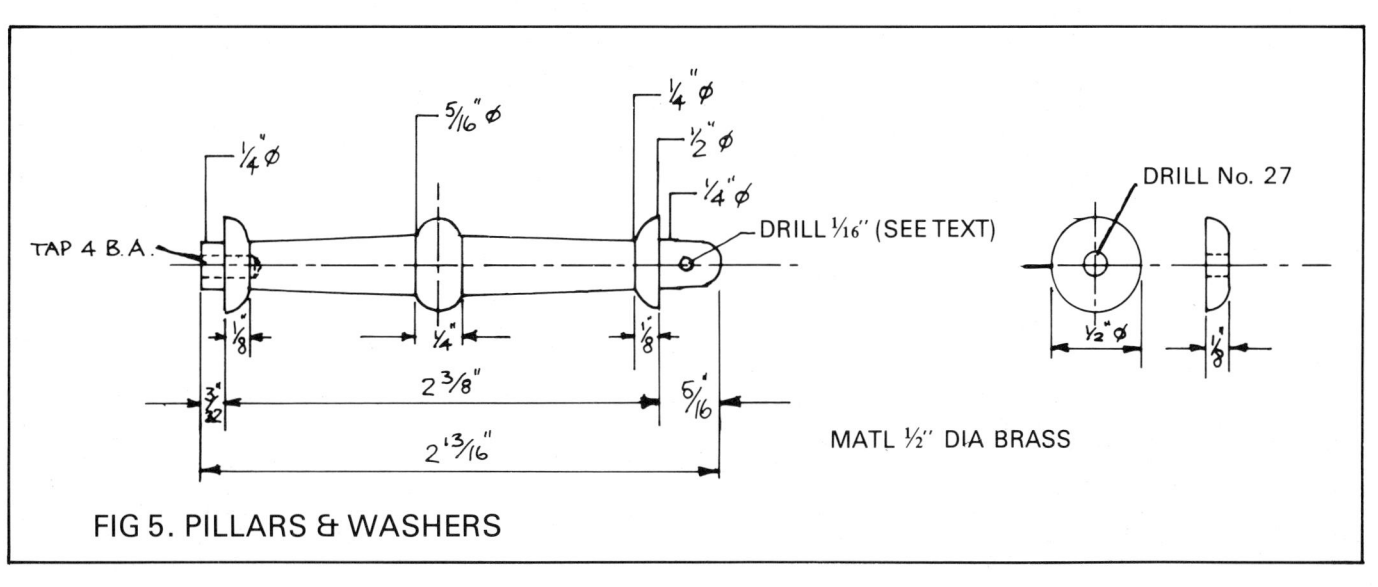

**FIG 5. PILLARS & WASHERS**

TAP 4 B.A.
DRILL 1/16" (SEE TEXT)
DRILL No. 27
MATL ½" DIA BRASS

15

⅛" either side of the centre between shoulders. Holding the pillar as in plate 2, set the form tool up in the toolpost, and line the tip of the tool up with your marking out. If you have made your form tools up to the drawing, feed the tool (1) in until it just touches the work, and zero the cross slide dial. Then slowly feed it in 0.125", which should leave a slight flat adjacent to the

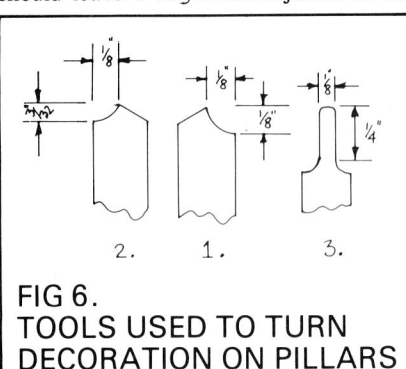

FIG 6.
TOOLS USED TO TURN DECORATION ON PILLARS

shoulder. Do this to both ends of each pillar. Set up form tool (2), and hold the job as in plate 3. Again, line up the tip of the tool with your marking out (remembering that this time the radius will be formed from the mark, towards the centre of the pillar), and feed the tool in so that you leave a small flat at what will become the centre of the bulge. Reverse in the collet and form the other half in exactly the same way, leaving the centre a full ½" dia. Your pillars should now look like plate 4. The final stage is to turn away the waste to taper the pillars as in the drawing. Grind up a tool as in Fig. 6 (3), set it up in the tool post, and set the topslide over to turn taper towards the headstock. In my case, a two degree taper was just right but you may need to experiment a little. The next operation demonstrates the use of one of my ½" dia. mandrels with a threaded end. This mandrel has a 4 B.A. thread, ⅜" long, and is held in the ½" collet. The short spigot is too short to hold firmly without damaging it or straining a ¼" collet, so screw the pillar onto the mandrel by its 4 B.A. thread, supporting the other end with a revolving centre (see Plate 5). Lock the saddle and taper turn from the central bulge to the shoulder radius. The taper should blend in nicely with the roots of the radii already formed. When you have done them all, reverse and hold by the long spigot and back centre, repeating the tapering on the other half of the pillars. Whilst the pillar is set up, remove any tool marks with emery sticks.

The last operation on the pillars is to cross drill the spigots and remove the unwanted centre drilling. The cross drilled holes are to take taper pins which will hold the front plate in position. As the pins should pull the plate down hard onto the shoulders of the pillars, the distance they are drilled away from the shoulders is

*Plate 2. Forming an end radius on the pillars.*

*Plate 3. Forming central bulge on pillars.*

*Plate 5. Taper turning pillar held on threaded mandrel.*

*Plate 4. Pillar with all forming done before taper turning.*

critical, so it is prudent to make a small jig to do the job. I made this from a bit of scrap ¾" square mild steel about 1" long as in the sketch (Fig. 7). Face one end in the four jaw chuck and mark out its centre. Carry this marking accurately round to one face of the block using a scribing block. Centre in the 4 jaw, drill, and ream ¼" right through. Now drill through ¹⁄₁₆" on the centre line, ⁵⁄₃₂" in from the faced end, to break through into

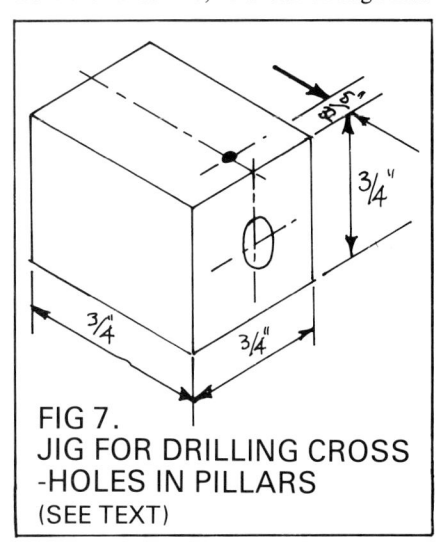

**FIG 7.**
**JIG FOR DRILLING CROSS-HOLES IN PILLARS**
(SEE TEXT)

the ¼" hole. Turn up a ¼" spigot on a piece of scrap brass, hold in the jig with the shoulder hard against the faced end of the jig, and drill through the spigot. Try fitting in one of the plates to see if pushing a taper pin in pulls the plate and pillar together. If all is OK, go ahead and drill all the pillars. If things are slack when your pin is fully home, take a light facing cut off the end of the jig and try again with a test piece until you've got it right. Plate 6 shows the set up with one of the pillars being drilled. When drilling is completed, we need to saw off the centre drilling, leaving the spigot ⁵⁄₁₆" long before tapering and rounding the ends. Hold the pillars in the vice fitted with fibre jaws for the sawing operation, and then transfer to the lathe to finish off. You will remember that when we were forming the radii earlier, I said leave slight flats at the shoulders and in the centre of each pillar, the full ½" diameter. The Myford ½" collet will just grip on these flats whilst we put a slight taper on the long spigot which will aid assembly later. About two degrees taper is sufficient (the angle is not critical), but leave about ³⁄₃₂" from the shoulder parallel at ¼" dia. Finally, with a file, radius the ends to finish off neatly. Polish out file marks with the emery stick.

To asemble the pillars and plates, we need four brass washers ½" dia. drilled No. 27 (4 B.A. clearance). Hold a piece of ½" dia. rod in the chuck, face the end and centre drill, then drill No. 27, about 1" deep. Using odd-legs, mark off ⅛" and radius the corner using the same form tool used on the centre of the pillars. Polish with emery sticks and part off ⅛" thick.

You will also need four 4 B.A. cheesehead screws, ½" long. The heads on standard screws are not well finished, so I modify the shape in the lathe. Hold a scrap of mild steel rod in the chuck, drill and tap it 4 B.A., ¾" deep. Screw in one of your cheeseheads and clean up the O.D. of the head with light cuts, then take a light facing cut across the slotted face. Slightly dome the slot face with a file and highly polish the head all over with emery sticks. Remove the screw, but take care not to touch the

*Plate 6. Cross-drilling pillars in jig.*

polished head with your fingers. I like to see blued screws in clocks, and this is easily done. Drill a clearance hole in a piece of scrap brass and place a screw in the hole. Hold over a gentle flame until the head turns a nice bluey-purple and quench in oil. This leaves a beautiful blue shiny head which looks well against polished brass.

Assembly of the plates is straightforward. Fit the short spigots into the backplate and fix with washers and screws. Before tightening the screws, make sure that the cross-drillings in the front spigots are all horizontal. Lie the backplate flat on the bench and the front plate should fit snugly over the spigots, with just a slight push needed to press it home to the shoulders. Fit taper pins into the holes and gently tap them home. The plate should now fit tightly onto the pillars with no sign of shake. Measure the gap between the plates in several places to make sure they are parallel. Plate 7 shows progress so far.

*Plate 7. The pillars and plates assembled.*

# Chapter 3

## THE BARRELS AND BARREL ARBORS

The barrel arbors are machined from ½" diameter silver steel. You will need two pieces each 5" long. As I have said before, I do as much lathe work as possible using collets to ensure concentricity and for ease of working, so I design pieces with standard collet sizes wherever possible. For these arbors you will need ¼", 5/16" and ½" collets, or you can do all the turning between centres, according to normal practice. Whichever way you decide to use, face both ends of the pieces of silver steel and centre drill with a BS1 centre drill. If you are working between centres, turn the arbors to the dimensions shown in Fig. 8, not forgetting to allow for the removal of the centre holes when the lathe work is finished.

Using collets, hold the rod in a ½" collet with the free end supported by a back centre and allowing enough extra length to remove the centre hole on completion, turn the 5/16" dimension up to the curved shoulder. Then turn to ¼" diameter, leaving 1/8" at 5/16" diameter. Form the radius using the same tool you used to form the ends of the pillars. The ¼" diameter section will be the pivot which runs in the front plate, in a ¼" reamed hole. The pivots should be highly polished and a nice running fit in their holes. To achieve this, turn the ¼" diameter to a fine finish and a tight fit in a ¼" reamed hole in a scrap of brass plate. Then polish using a fine emery stick until the pivot is a nice fit in the hole. Make sure that the shoulder which will butt up to the inside of the plate is perfectly square with no radius in the corner which may cause binding. Special double ended pivot file/burnishers are available for polishing pivots if you wish to buy one, (see suppliers list), and these make a superb job of pivot polishing.

The next stage is to hold the arbor by the ¼" diameter, again with the free end held by a centre. Turn the long 5/16" length, leaving the ½" diameter section 1/8" thick as in the drawing. Leave the 5/16" dimension a little full so that the arbor will be a tight fit into the barrel flanges. Turn the short ¼" diameter pivot and polish to a nice fit as already described. The 2⅜" dimension in the drawing will need to be modified to suit the dimension between your assembled plates. When the arbor is assembled in the plates later, there should be a small amount of endshake (about 1/64"). If your plates are exactly parrallel, as they should be, you can check the shake by laying the arbor across the edges of the assembled pillars and plates, and adjust accordingly. Do not turn the groove for the slip washer yet as its location will depend on the finished sizes of the barrels and great wheels.

There are two methods of forming the winding squares accurately. If you have a vertical milling machine and a method of simple dividing or a dividing head, then this is the obvious method to use. The square should be slightly tapered towards the end of the arbor as this makes it easier to insert the key for winding. I set up my home made dividing head on the milling machine table with a piece of tinplate packing under the collet end of its base. This lifts the free end of the arbor slightly and produces about the correct degree of taper on the finished square. Mark out the length of the square on the ¼" diameter section, and hold in the dividing head by the long 5/16" section, preferably in a collet. I use a ⅜" end mill to cut the square. Feed the cutter down and take about a 0.020" cut across the rod. Index round 90 degrees, which on my 60:1 dividing head equals 15 turns of the handle, and take another cut across. This should leave you with two flats at right angles with a slight radius at the corner between them. Feed the cutter down slightly so that a second cut across one face will halve the

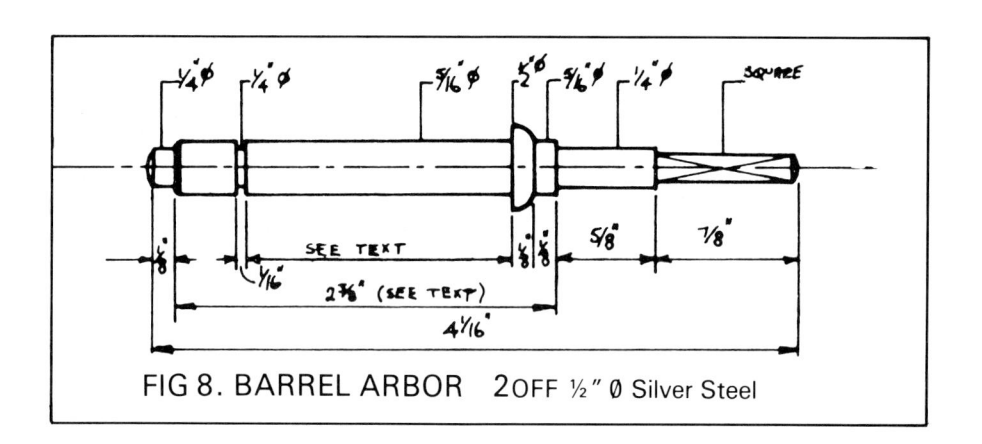

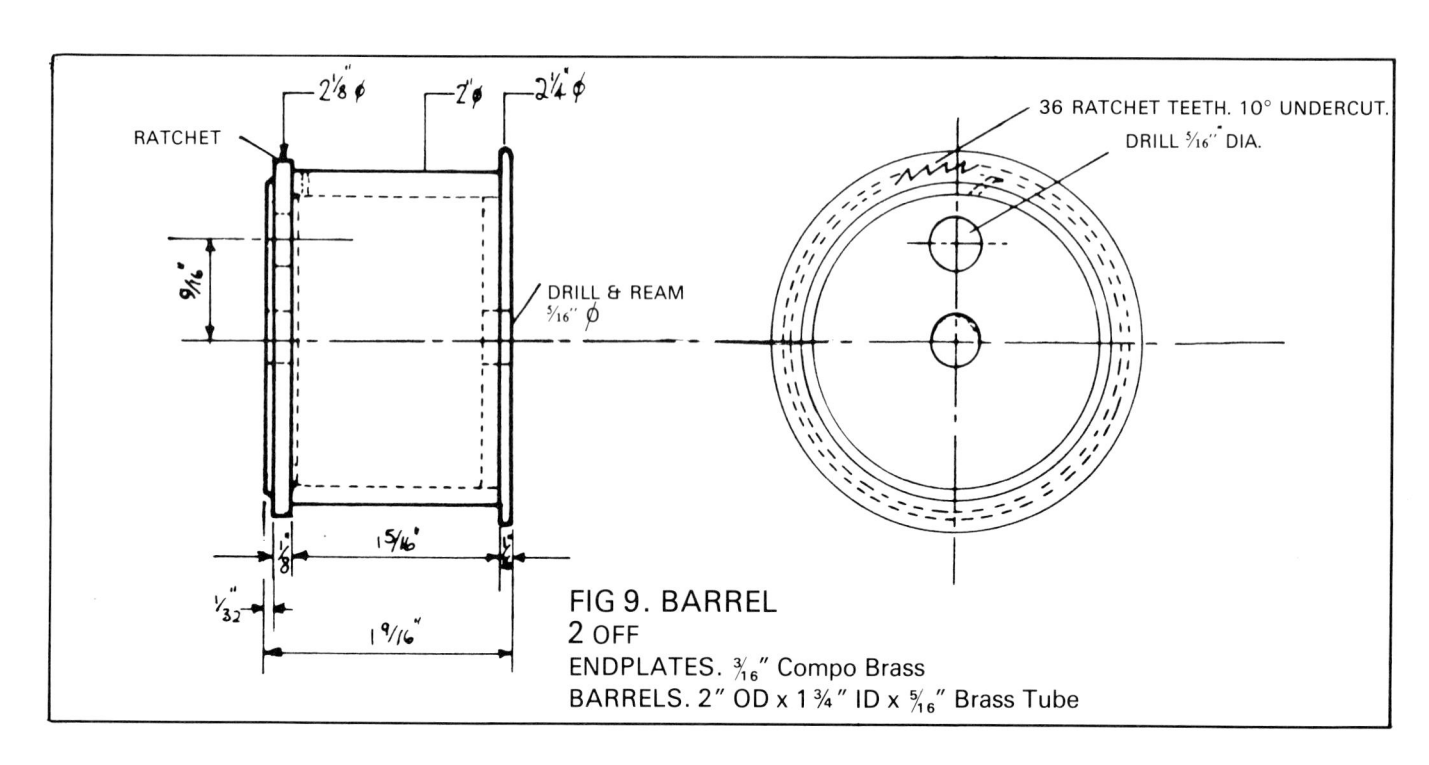

**FIG 8. BARREL ARBOR** 2 OFF ½" Ø Silver Steel

**FIG 9. BARREL**
2 OFF
ENDPLATES. 3/16" Compo Brass
BARRELS. 2" OD x 1¾" ID x 5/16" Brass Tube

*Plate 8. Machining Barrel end plate to fit tube.*

radius that is left between the faces and then index back to the first face where a second cut across should leave a nice sharp corner. When the tool is set correctly, mill off each face of the square in turn, working to a table stop set to give the correct shoulder length. The second arbor can then be machined at the same setting. When the squares are finished, saw off the centered end of the arbor to the correct length and slightly radius the end in the lathe. Drawfile the faces of the squares and polish with emery sticks.

If you haven't got the facilities to mill the squares, they can be filed quite accurately using a simple home made roller filing rest in the lathe. The work is held in the hedstock in a collet if possible, and is indexed using a simple attachment which has been described many times in the past. A suitable lathe changewheel which is divisible by 4 eg. (20, 40, 60), is fixed into the back of the hollow headstock mandrel using an expanding plug which grips inside the mandrel like a 'Rawlbolt'. A spring or screw detent can then be fixed to any convenient part of the lathe so that it can engage the gear teeth on the wheel and thus lock the mandrel precisely in the required positon. Using a 40 tooth wheel, engaging the detent every tenth tooth will give four equal divisions.

Roller filing rests are very useful lathe accessories which are well worth buying (see suppliers list) or can be made quite easily. The basic requirement is a bracket which can be fitted to the cross-slide, vertical slide or 4 tool turret. The bracket carries an angle iron frame with screw height adjustment. Mounted on the vertical face of the angle are two steel rollers about ½" diameter, which are free to revolve on pins. In use, the attachment is fitted up with the rollers parallel to the lathe bed, with one roller in front and one behind the workpiece. The height of the rollers is then adjusted until a file will cut the workpiece. File down until the file rubs on both rollers and ceases to cut, index round 90 degrees and repeat the filing operation. Adjust the roller height again until a sharp corner is formed between the two faces, then continue to file all 4 sides. If you turn the rollers and leave a shoulder on one end, this will act as a length stop when you use the attachment. As I said earlier, filing rests are well worth having, they are very useful for filing all sorts of shapes and flats on round bars. One use that immediately springs to mind is filing D Bits accurately to half diameter.

With the winding squares done, we can put the arbors to one side, leaving the groove until the barrels etc., are finished. We need to use the centre hole, which is still left at the opposite end to the square, when we cut the teeth on the barrel ratchet so leave it there until later.

For the barrels you will need two pieces of drawn brass tube, 2" outside diamter x 1¾" inside diameter x 1⅜" long. The endplates are cut from 3/16" compo brass. This can be obtained from the suppliers as circular blanks cut from sheet, and you need 4 blanks 2¼" diameter x 3/16" thick CZ 120 brass. I buy this in sheet form and saw my blanks with a bandsaw which is a cheaper method if you use a lot. However, buying standard blanks can save a lot of work if you haven't got a bandsaw!

Grip the tube in the three jaw chuck and, taking light cuts, face both ends to a finished length of 1 5/16". Remove any burrs and measure the wall thickness of the tubes at two or three places at both ends. I usually find that the bores are not concentric to the outside diameter. Using a small boring bar, take light cuts off the inside of the tubes until they are just cleaned up all over. Put the tubes to one side for now until the end plates are finished.

If you have bought blanks for the endplates, hold each one in turn in the three jaw chuck, centre drill, drill and ream 5/16". If you saw the blanks from sheet you should have a centre punch mark in the middle. Drill in the drilling machine and again, ream all four centre holes 5/16". You now need a mandrel made from ½" diameter mild steel. A piece about two inches long is held in a collet or 3 jaw chuck, faced and turned to a good fit in the 5/16" holes for a length of about ½". Turn this 5/16" spigot down again to ¼" diameter leaving 5/32" at the shoulder. Thread the spigot ¼" BSF. You can now fit an endplate blank onto the 5/16" part of the mandrel and fix it to the mandrel using a ¼" BSF nut, not forgetting

*Plate 9. Grooving the barrels.*

a washer to prevent marking the brass. Held in this way all turning on the endplates is bound to be concentric with the centre hole.

The two blanks which form the ratchet ends of the barrels are turned to a diameter of 2⅛" and are then turned down to leave a spigot of 1/32" long which is a good fit in the inside of the barrel tube. They are then reversed on the mandrel and using a lathe tool with a small radius on its tip, take about 1/32" off the face of the blank leaving a raised ring with an outside diameter of 1⅛" and a width of 1/16", and a ½" diameter boss in the centre of the blanks. This is done only to reduce the contact area between the barrels and great wheels which reduces friction whilst winding. Details are shown on the drawing (Fig. 9). The opposite endplates are made in the same way except that they are 2¼" diameter and the spigots which fit into the barrels are ⅛" long. This leaves a flange 1/16" wide and the outside diameter of this can be given a slight radius, as in the drawing. The faces of these last endplates do not of course need relieving. Plate 8 shows one of the endplates being machined. The additional 5/16" holes in the ratchet ends are used to feed the end of the lines through after tying stopknots.

We now come to an optional operation grooving the barrels. This is done to help the lines to wind onto the barrels evenly. I usually machine these two grooves but if the clock is set up perfectly level as it should be, they aren't absolutely necessary. Most antique clocks had them but some didn't, so it's up to you.

The set up I use to do the grooves is shown in Plate 9. The first thing to do is to fit the ratchet endplates to the barrels, using silver solder. Make sure that the mating faces are clean and apply 'Easyflo' flux mixed to a creamy consistency with water. Push the barrel tube onto the endplate and stand the assembly on end on a firebrick with the open end of the tube upwards. Put some more flux round the joint inside the barrel and warm things up evenly all round. I use a Sievert propane torch and it doesn't take long to heat up. I've also made a small rotating stand for soldering etc., rather like a potters modelling stand, which makes it easy to revolve the barrels whilst heating. Run some easy silver solder round the inside and watch for it to appear all round outside. Let the job cool and pickle in a 10% sulphuric acid solution. Keep a separate pickle bath just for brass otherwise everything gets a coating of copper which is a nuisance.

To hold the barrels in the lathe, fit them back on the mandrel used for turning the endplates and make a mild steel disc about ¼" thick, the same diameter as the barrel tube, with a spigot which fits the tube bore nicely. The free end of the tube can then be supported by a back centre, without a flange in the way of the tool during the grooving operation.

The lathe tool needs to be about 3/32" wide with a semi-circular end, easily ground up from a piece of high speed steel. Take a light cut along the length of the tube to make sure it's true, and then set the lathe up to cut 14 threads per inch. You can cut the groove using a slow backgear speed but make sure you don't let the tool run into the ratchet flange. If you aren't too confident about this, it may be best for you to make a handle to fit the back end of the headstock mandrel and turn the work by hand. Take light cuts, finishing when the groove is about 1/32" deep. Don't worry about a little chatter as the tool cuts, as any marks can be removed with a round needle file.

With the grooves finished, we can now

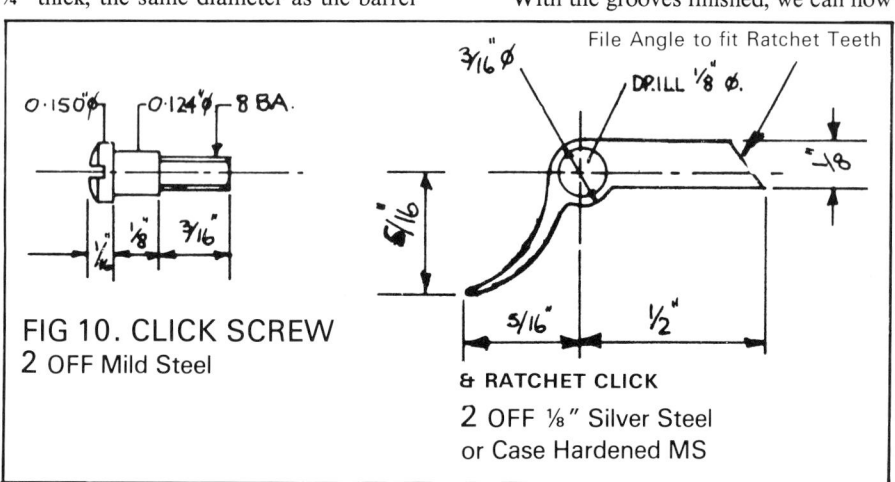

**FIG 10. CLICK SCREW**
2 OFF Mild Steel

**& RATCHET CLICK**
2 OFF ⅛" Silver Steel
or Case Hardened MS

finish the assembly of the barrels and fix them to their arbors. The easiest method is to soft solder the remaining flanges to the tubes and then soft solder the barrel assemblies to their arbors. Clean the end faces of both barrels and about ¼" inside each tube, as well as both mating surfaces on the flanges. Flux the barrel ends, heat them up and coat the joint areas carefully with soft solder. Do this carefully to avoid getting solder on the outer surfaces. Place the flanges in position and heat evenly until the solder melts and the flanges slide into place. Make sure they are fully home before the joints cool.

To fit the arbors, flux round the ½" diameter shoulders and tin with solder. Flux the holes in the barrels after ensuring that they are thoroughly clean and slide the barrels onto their arbors, 2¼" diameter ends first, gently heating the assembly until the solder melts and the arbors slide fully home so that the shoulders fit tight against the 2¼" diameter flanges. Quickly heat the ratchet ends and run a little solder round the arbors until it flows into the joints. Use as little solder as you can get away with to avoid too much cleaning up. If you do the job properly, there should be a tiny fillet of soft solder round the arbors at each end of the barrels.

The last major operation on the barrel assemblies is to cut the ratchet teeth. This will be dealt with when I describe wheel cutting, as I find it easiest to set up and cut all the wheels at one setting.

## RATCHET CLICKS, SCREWS AND SPRINGS

The ratchet clicks are mounted on pivot screws on the rims of the great wheels and are held in mesh with the ratchet teeth by curved brass springs also mounted on the great wheels. These parts are shown in Figs. 10 and 11, and are quite straightforward to make. Start with the click. This part always shows a lot of wear on antique clocks and often needs replacing. The most common type of click was made in one piece with a threaded pivot, so that when the click is screwed loosely into the great wheel the thread itself acts as the bearing surface. This design invariably leads to sloppy action due to wear in the threaded holes, followed by attempts to repair the problem by punching the holes smaller. The method used for this clock, with a pivot screwed firmly to the wheel, upon which the click can revolve freely, is much less prone to wear.

Mark out the clicks on a piece of mild steel or silver steel. I prefer silver steel as this is easy to harden and temper, but case hardened mild steel will do fine. Allow enough metal to leave the acting angled end of the click about ¹⁄₁₆" longer than on the drawing so that this can be filed to fit the ratchet teeth after a trial assembly. The curved leg aids assembly and final setting up when the clock is finished, as it makes it a simple job to disengage the ratchets and thus allow the barrels to revolve freely whilst sorting out the lines. The degree of curvature is not vital but stick to the overall dimensions otherwise you may have problems with clearance when you assemble the great wheels and barrels.

Drill and ream the pivot holes ⅛" diameter before cutting out and filing to shape, not forgetting to leave the extra length on the acting face. Drawfile the edges using needle files and polish all surfaces with emery sticks. Later, after fitting the click, harden and then temper to a dark straw colour if using silver steel, or case harden with 'Kasenit' or a similar compound if using mild steel.

The pivot screw is a simple turning job; the only points to watch are that the click is a good "running fit" without too much shake and that the ⅛" long bearing surface is a fraction longer than the thickness of the click, so that the click moves freely when the screw is fully tightened, again without too much shake. Part off the screw leaving the head on the thick side and saw the slot using a junior hacksaw. Return the screw to the lathe, face the head to the correct thickness and using a fine file, slightly dome the head. Finish off by polishing with fine emery sticks and blueing as you did with the plate screws.

The click springs are again quite straightforward sawing and filing jobs. Mark them out as in the drawing Fig. 11, and cut out using a piercing saw. I have a Cowells Jig Saw and find it invaluable for jobs like this. It cuts thin brass easily and cleanly and is a good deal quicker than a piercing saw. Drill the holes for the fixing screws and steady pins, (tapered clock pins), and finish with fine files and emery sticks. When we have mounted the ratchets to the great wheels we will have to bend the springs to match the rims of the wheels. This bending will harden the springs sufficiently to produce a nice action. Rough stamped click springs are available from clock material suppliers ready to be filed to shape and finished. These are useful if you are making several clocks as they save a lot of time sawing, but making just two springs is not a long job.

## SLIP WASHERS

These are shown in Fig. 12 and are used to hold the great wheels in position on their arbors. Again, they are easy to make and there are two methods of producing them. Either mark them out on ¹⁄₁₆" thick brass sheet and drill, saw and file to shape, or, as I do, make them from a piece of 1" diameter brass bar. To do this, face one end and centre punch the middle. Scribe a line across the diameter and mark out the hole positions. Drill the ⁵⁄₁₆" and No. 43 holes with the bar held vertically on the drilling machine and then return to the lathe; centre drill and drill ¼" diameter. All these holes should be made about ½" deep. Take a light cut along the bar to clean up the outside diameter, and then part off two ¹⁄₁₆" thick slices. Countersink the No. 43 holes to take 8BA countersunk screws. File the slots between the ¼" and ⁵⁄₁₆" diameter holes and finish by polishing all over.

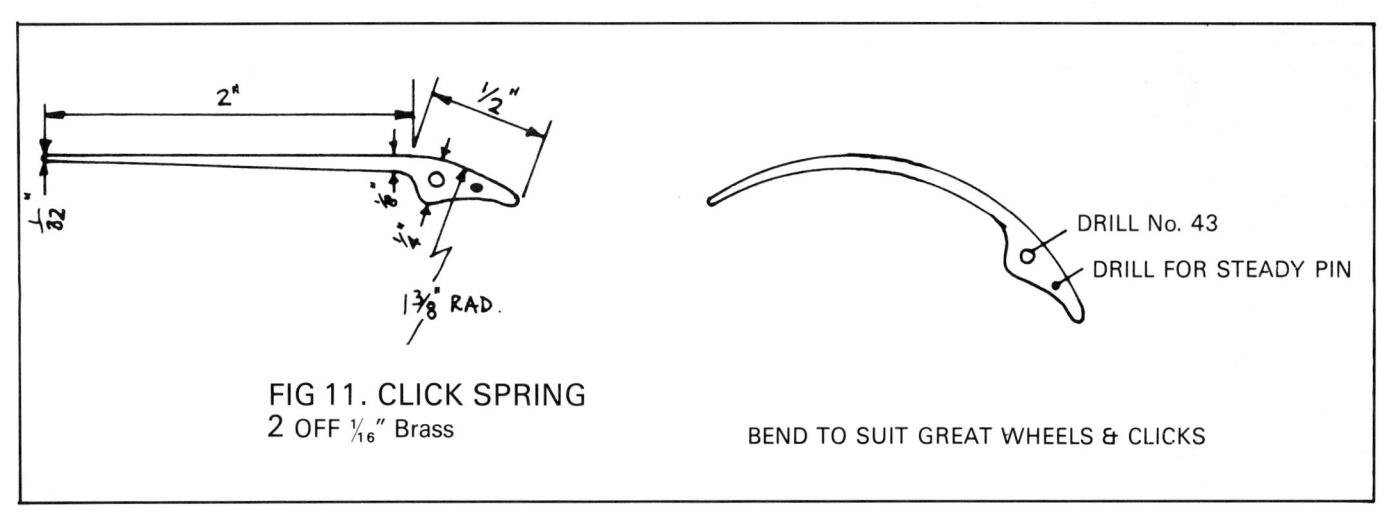

**FIG 11. CLICK SPRING**
2 OFF ¹⁄₁₆" Brass

BEND TO SUIT GREAT WHEELS & CLICKS

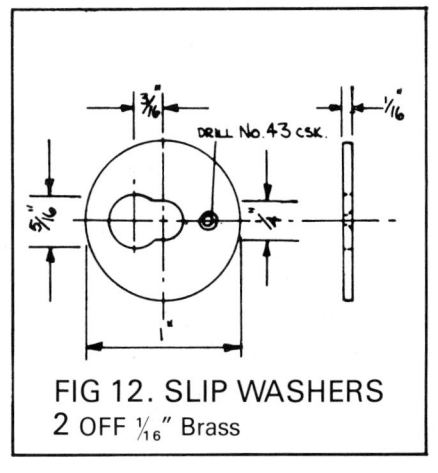

**FIG 12. SLIP WASHERS**
2 OFF ¹⁄₁₆" Brass

# Chapter 4

**CUTTING THE BARREL RATCHET TEETH.**

After reading the section on gearcutting (see Appendix 1) you should, I hope be able to sort out a method to suit your workshop conditions, so when you have equipped yourself for the job you can start by cutting the barrel ratchets.

The ratchets are cut on the ⅛" thick flange of the barrels. Fig. 12. shows the shape of the flycutter. This can be filed up from a piece of silver steel, or better still, ground on the end of a piece of ¼" dia. H.S.S. Don't forget the clearance angles. The teeth should be undercut 10 degrees, and this is achieved by offsetting the cutter from the barrel centre line. This undercutting is done so that the ratchet pawl 'digs in' to the roots of the teeth positively after winding. The amount of offset can be worked out mathematically and for an undercut of 10 degrees on a wheel of 2⅛" dia, the offset should be 0.184". There is a much easier practical method of setting up offset. To do this, start with the tip of the cutter on the blank centre line and cut a tiny

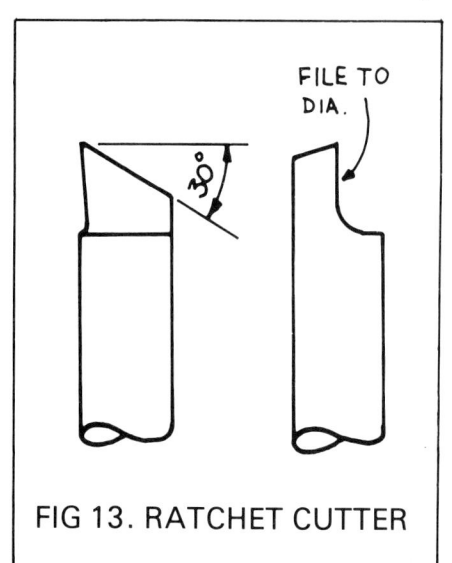

*Plate 10. Cutting the ratchet teeth.*

**FIG 13. RATCHET CUTTER**

line on the rim of the work. Then rotate the blank one tooth (for 36 teeth, 1 tooth – 10 degrees). If the cutter is now offset until it's tip is exactly in line with the line just cut, you will get the correct amount of undercut when you cut the teeth. It is most important that you cut the teeth the right way round and that you offset the cutter the right way, so I have included a photograph of my set up for cutting the ratchets on the milling machine. The barrel is held by it's arbor in a ¼" collet, with the ratchet end supported by a centre. (See Plate 10). In the photo the cutter is offset to the right of the barrel centre line. The picture also shows the cutter holder I use.

When all is ready, cut the first tooth, but cut it too shallow. Index round the next tooth and cut again. Now adjust the depth of cut until the tip of the tooth is not quite a sharp point, but has a tiny flat just a few thou wide. If you have made the cutter properly, the root of the teeth should be just above the surface of the barrel. If this is so, go ahead and cut all the teeth. If the cutter touches the barrel, regrind it to a slightly shallower angle and try again.

**THE GREAT WHEELS.**

To complete the barrel arbor assemblies we need to make the Great Wheels and machine the grooves to take the slip wasners.

In the introduction, I said that I hoped to use the same Module cutter throughout, but this brought a slight problem with the strike

*Plate 11. Cutting wheels on the milling machine.*

train Great Wheel. An 84 tooth wheel of 0.8 Module should be 2.730" dia. If we make it this size, there is hardly enough room for the ratchet click to be mounted on the rim of the wheel. In theory, it would be better to use a 0.85 Mod. cutter for this wheel, which gives us a diameter of 2.901". It is interesting to note that a 0.85 Mod. cutter is only 0.0025" wider than 0.8 Mod. but that this tiny amount makes a difference of 0.171" on diameter. On examining the wheel from the antique clock, which is near enough 2.9" dia. as makes no difference, I found that a 0.8 Mod. cutter fitted the tooth space perfectly. This is a typical case of the latitude found in old clocks which run

*Plate 12. The finished barrels and great wheels.*

perfectly well. The maker used a larger blank which gave him slightly thicker teeth. I have tried both 0.8M and 0.85M on a blank 2.9" dia. with 84 teeth and both meshed well in the depthing tool with the slightest adjustment which is of no consequence, although the 0.85 Mod. wheel is technically correct. If you wish, make a 0.85M flycutter as described elsewhere, to cut a space 0.0525" wide. The alternative method is to cut all the teeth with a 0.8 Mod. cutter, then index the wheel round a fraction of the tooth space and take a second cut round to produce teeth the same width as the spaces. This latter method is perfectly acceptable. Indeed, on wheelcutting engines in production today, provision is made for fine adjustment of the detent to cope with this process, which saves buying or making the occasional odd cutter. Apart from this one case, all other wheels are 0.8 Module.

Cut the blanks from 3/16" compo brass, slightly oversize. The finished sizes are: going train great wheel 3.108" dia.; strike train 2.900" dia. Drill and ream centre holes 5/16" dia. Using the mandrel made to turn the barrel endplates, mount each in the lathe and turn to diameter. Plate 11 shows my set up for wheelcutting, using a multi tooth cutter mounted on a stub arbor in the miller, with the blank held on a mandrel in the dividing head. (See section on gearcutting). Set up to cut 84 teeth on the smaller blank and centre the cutter. Cut the first two teeth too shallow and adjust the depth of cut until the tips of the teeth are properly formed. Then cut all round and repeat the process but with 96 teeth on the larger blank.

You can continue with all the wheels if you want at this stage, but I like to get the barrel assemblies finished as this gives a feeling of progress! With all the teeth cut, slide the great wheels onto their arbors and scribe a mark on the arbors to position the slip washer grooves. Hold the squared end of the arbors by the ¼" dia. section in a collet. Support the other end of the arbor with a tailstock centre and using a 1/16" wide parting tool, machine the grooves as in the drawing Plate 8. Try the slip washers with the great wheels in place. If they are tight, check if the grooves are deep enough and wide enough, or thin the washers down slightly by rubbing them on a sheet of emery cloth. If they are loose, the washers can be dished slightly. The aim is to get the wheels to revolve freely, but without any wobble, on the arbors. The washers are held in place with small countersunk screws through into the wheels, or simply by drilling through the washers into the wheels and tapping a clock pin through both, but it is best to leave this until the wheels are crossed out.

Fig 14 shows the Great Wheels. To mark out the crossing, coat one face of the wheels with Spectra Layout Blue, and make a small plug to fit the centre holes with it's centre marked with a punch. Fit this plug

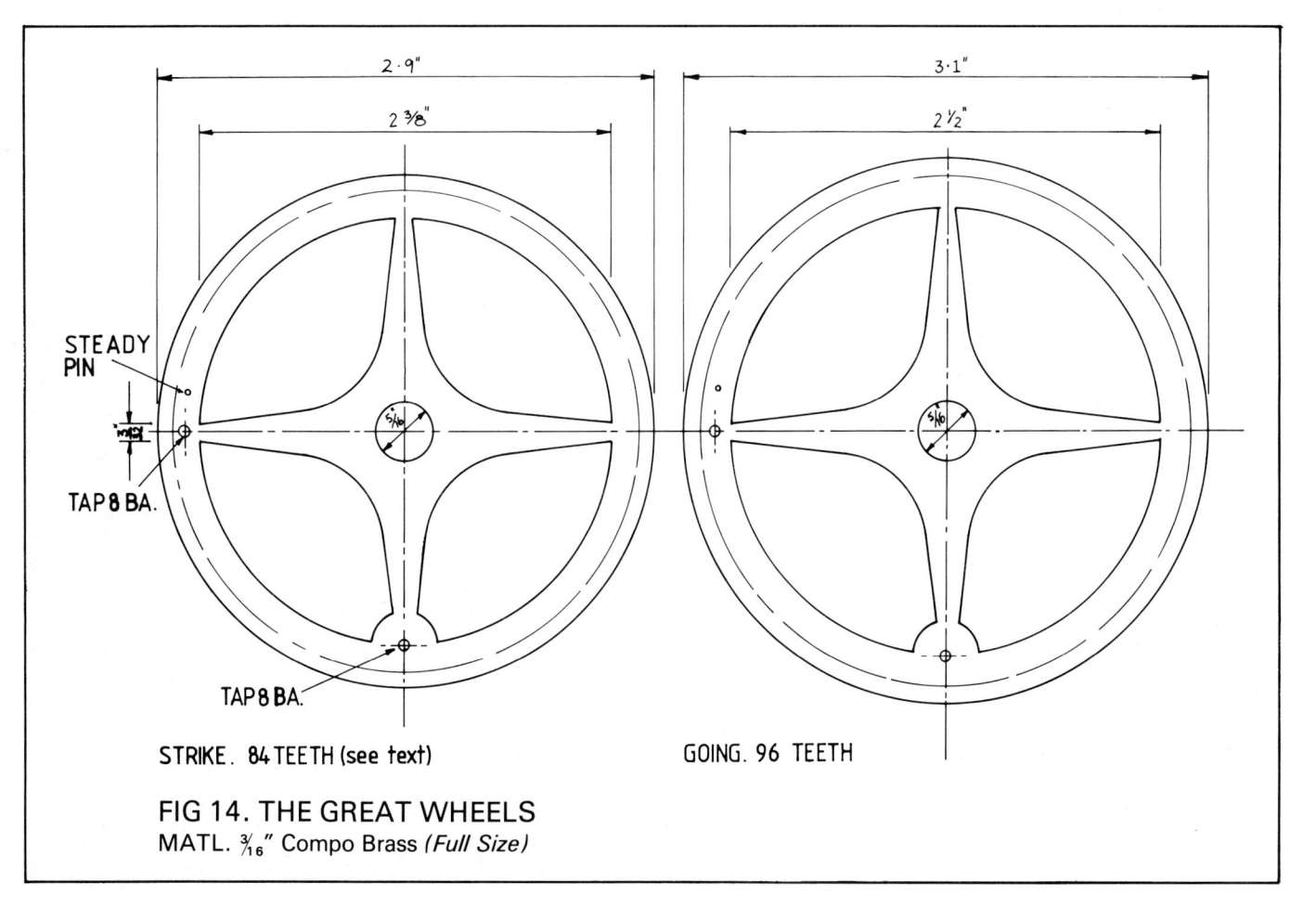

STRIKE. 84 TEETH (see text)            GOING. 96 TEETH

**FIG 14. THE GREAT WHEELS**
MATL. 3/16" Compo Brass *(Full Size)*

into the wheels and mark out the circles with dividers. Scribe two lines at right-angles to each other across the diameters. This can be done using a scribing block and revolving the job 90 degrees in the dividing head, or you can scribe the first line across the tips of two diametrically opposite teeth, then count round a quarter of the number of teeth and scribe the second line across. It is best to make a template of the spoke shape and use this to mark out the crossings. Thin card will do for this or you can make a more durable pattern from tin plate. The boss which carries the clicks is marked out with dividers.

When the marking out is complete, drill a small hole in every corner where a spoke meets the rim. The waste metal is removed with a piercing saw. If you have a vertical mill and a rotary table the circular rim can be machined, which makes life easier and ensures a perfectly concentric rim, but the saw doesn't take long once you get the rhythm. The sawn surfaces are cleaned up with crossing and needle files. I will give more details of crossing out later.

The clicks and springs are now fitted as in the drawing, and the ends of the clicks are carefully filed to fit the ratchet teeth properly. Plate 12 shows the barrel assemblies. A click and spring are shown, illustrating how the click should engage the teeth and be drawn fully into the roots of the teeth. You can also see how the end of the spring is bent to give a good positive action to the click. At this stage, the barrels are ready to be 'planted' between the plates.

## WHEELS

I find it convenient to cut all the wheels at one setting as this saves a lot of time setting up etc. All the remaining wheels are cut from 1/16" Compo Brass, with the exception of the two 39 tooth motion wheels (minute and reverse minute), which are 3/32" thick. As I've mentioned before, I cut all my blanks from sheet stock using a bandsaw as this is the cheapest method and also saves having to order blanks. Circular blanks are available from suppliers in a wide range of stock sizes. Chronos Designs will supply a complete set of blanks for this clock, including barrel tubes and endplates. If you cut your own blanks, make them about 1/16" oversize to allow for machining to diameter. Table 1 lists the finished diameters, numbers of teeth, thickness and centre hole diameters of all the blanks needed.

With all the blanks cut, drill and ream all the centre holes. I have made all but one wheel with a 3/16" hole as this means that we only need to make a couple of mandrels, and all the wheel collets which mount the wheels to the arbors will be the same. This makes for much easier production. The mandrel for the 3/16" bore wheels is made on the end of a piece of 1/2" dia. mild steel rod, about 3" long. Hold the rod in a 1/2" collet and turn the end down for a length of about 1/2" so that it is a good tight fit in the 3/16" holes. Thread the end of the spigot 2 BA leaving 3/16" at the shoulder as a seat for the blanks. I leave this amount unthreaded so that there is plenty of room to mount the two minute wheels and the two wheels with 56 teeth so that I can turn them and cut the teeth in pairs, thus saving repetition, but don't forget that if you make a mistake you'll ruin both blanks! It is also a good idea to support the blanks with a large washer behind them as a support to prevent flex and vibration whilst cutting the teeth. I use old brass wheels that didn't turn out right as washers!

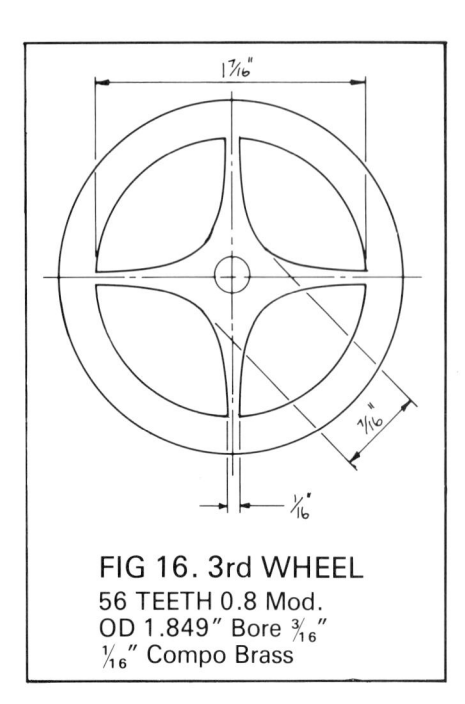

**FIG 16. 3rd WHEEL**
56 TEETH 0.8 Mod.
OD 1.849" Bore 3/16"
1/16" Compo Brass

slide which affects this setting so that it can't be moved by accident. You can then go ahead and cut all the blanks as previously described.

| Wheel | No. of Teeth | Outside Dia. | Thickness | Centre Hole |
|---|---|---|---|---|
| Centre | 60 | 1.975" | 1/16" | 3/16" |
| 3rd | 56 | 1.849" | 1/16" | 3/16" |
| 'Scape | 30 | 1.650" | 1/16" | 3/16" |
| Pin | 56 | 1.849" | 1/16" | 3/16" |
| Pallet | 49 | 1.628" | 1/16" | 3/16" |
| Warn | 42 | 1.408" | 1/16" | 3/16" |
| Hour | 72 | 2.353" | 1/16" | 1/2" |
| Minute (2 off) | 39 | 1.313" | 3/32" | 3/16" |

**WHEEL BLANK DETAILS**

With the mandrel held in the lathe in a collet, mount each blank in turn, fixing the blanks to the mandrel with a 2 BA nut and washer, and machine all the blanks to diameter. The sizes given are calculated using the addenda for the cutters I used so check the calculations using the data for the cutters you propose to use. All the formulae needed have already been given and it is a simple matter to check. Don't forget that a variation in diameter of 0.010" or so is of little consequence when making a new clock as this will only affect the tooth thickness slightly and any small variation from the theoretical diameter can easily be accommodated when we plant the wheel trains between the plates. Making a replacement wheel for an old clock is a far different matter as we have to use the existing pivot holes so the wheel diameters are far more critical. (In many respects, it is easier to make a new clock than it is to repair an old one!).

With all the blanks the correct size, set up your wheelcutting arrangement using a 0.8 module cutter. Don't forget to centre the cutter exactly. I do this on my horizontal mill, holding the cutter on a mandrel held in an end mill holder. With the dividing head mounted on the machine table, I hold a lathe centre in the spindle and centre the cutter to the point. A piece of white paper held behind the cutter makes it easy to see clearly. With the cutter centered, lock the

The mandrel for the hour wheel is made from 3/4" dia. rod with a length at one end reduced to 1/2" to fit a collet. Hold the mandrel in the collet to turn the spigot to fit the hour wheel blank in the same way you did with the 3/16" mandrel, but this time you can use a bigger thread, say 1/4" B.S.F.

If you are using the lathe to cut your wheels, it pays to think ahead and make any mandrels, etc., that will be needed later, before setting up for wheelcutting. There is nothing more frustrating than getting everything set up and going smoothly, only to have to take it all down again to turn a forgotten mandrel! If all goes well, you should be able to prepare all the blanks and cut the teeth in an evening. I will deal with the escape wheel when I describe making the escapement.

## CROSSING OUT

In clockmaking, the spokes of the wheels are called crossings. Besides looking nice, crossing out wheels also lightens them considerably, thus reducing inertia in the wheel train. It is surprising how much weight is saved over the whole train. Remember that the wheels in the going train are stopping and starting every second, so if the inertia in the train is reduced to a minimum, efficiency will be increased. This results in a smaller driving weight, less

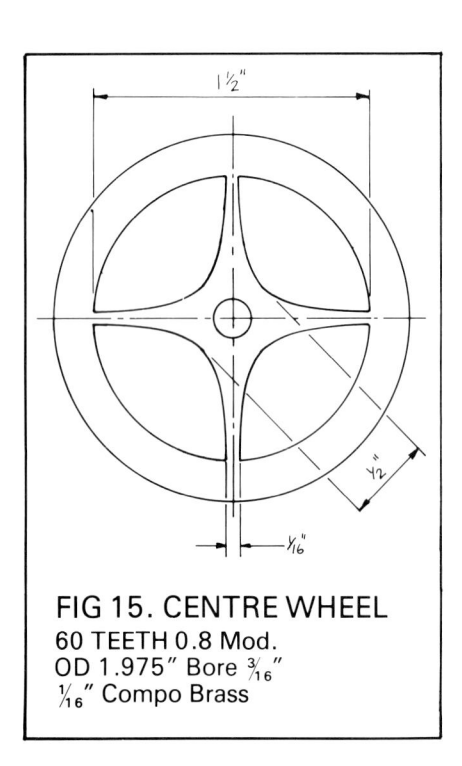

**FIG 15. CENTRE WHEEL**
60 TEETH 0.8 Mod.
OD 1.975" Bore 3/16"
1/16" Compo Brass

wear and more accurate timekeeping. There is no need to cross out the motion work. It is worth taking a lot of trouble with crossing out. Nicely proportioned, clean crossings with good sharp corners and smooth curves are a pleasure to see and a sign of good work.

To mark out the crossings, make a ³⁄₁₆" dia. plug to fit the centre holes from a piece of scrap, with a centre punch mark in its centre. Hold this vertically in the vice, with about ¹⁄₁₆" projecting above the jaws so that each blank can be placed on the plug, resting on the vice jaws. In this way, the rim of the crossing can be scribed using dividers from the centre of the plug. I usually coat one face of the wheel with layout blue to make marking out clearer. Next, scribe two diametrical lines across the wheel at right angles to each other. (A centre square will be found very useful). Another circle should also be scribed to mark the position of the bottom of the curves of the crossings. The diameters of these circles are given in the drawings. The next stage is to mark out the width of the 'spokes' where they meet rim so that you can insert the blade. Piercing saws should be used vertically, cutting on the down stroke. The work is usually held by hand, resting on a sawing board clamped to the bench or held in the vice. In this way, the work can easily be revolved to the best position. This method of holding can be tiring on the fingers if you are crossing out a few wheels and the blades are easily broken if the job twists or lifts. I often hold a toolmakers clamp in the vice so that it's jaws stick out of one side of the vice. I then clamp the wheel in the toolmakers clamp after inserting the blade and fitting it into the saw frame. In this way, the wheel is held firmly and effortlessly whilst cutting, but it can be moved around easily by simply slackening the clamp. Chronos Designs now market a very good piercing saw table with a work clamp.

If you saw carefully, you can cut almost exactly on the line, leaving very little to clean up with the file. With a little practice, you will soon be able to saw the crossings quickly and accurately. Pay particular attention to keeping the saw vertical in both a sore thumb so be extra careful. When you have got the shape right, drawfile with fine files, then wrap either very fine emery cloth or wet and dry paper round a needle file and repeat the process until all marks and scratches are removed. You can leave it at that or get an even better finish by burnishing. Tapered oval burnishers are available but steel knitting needles and sewing needles fitted into dowel handles are just as good. Burnishers are used just like drawfiling and produce a mirror finish, but the surface must be free of blemishes before you start burnishing.

When you have polished the crossings, rub the flat faces of the wheel on progressively finer emery cloth on a flat surface. I have a piece of ¼" plate glass for this job. Offcuts are easily obtained very cheaply from glass merchants and the sharp edge can be removed by rubbing with an oilstone. A piece of glass such as this also doubles as a cheap, accurate alternative to a surface plate! Finish the wheel faces on crocus paper, which is a very fine brown coloured abrasive paper which gives a

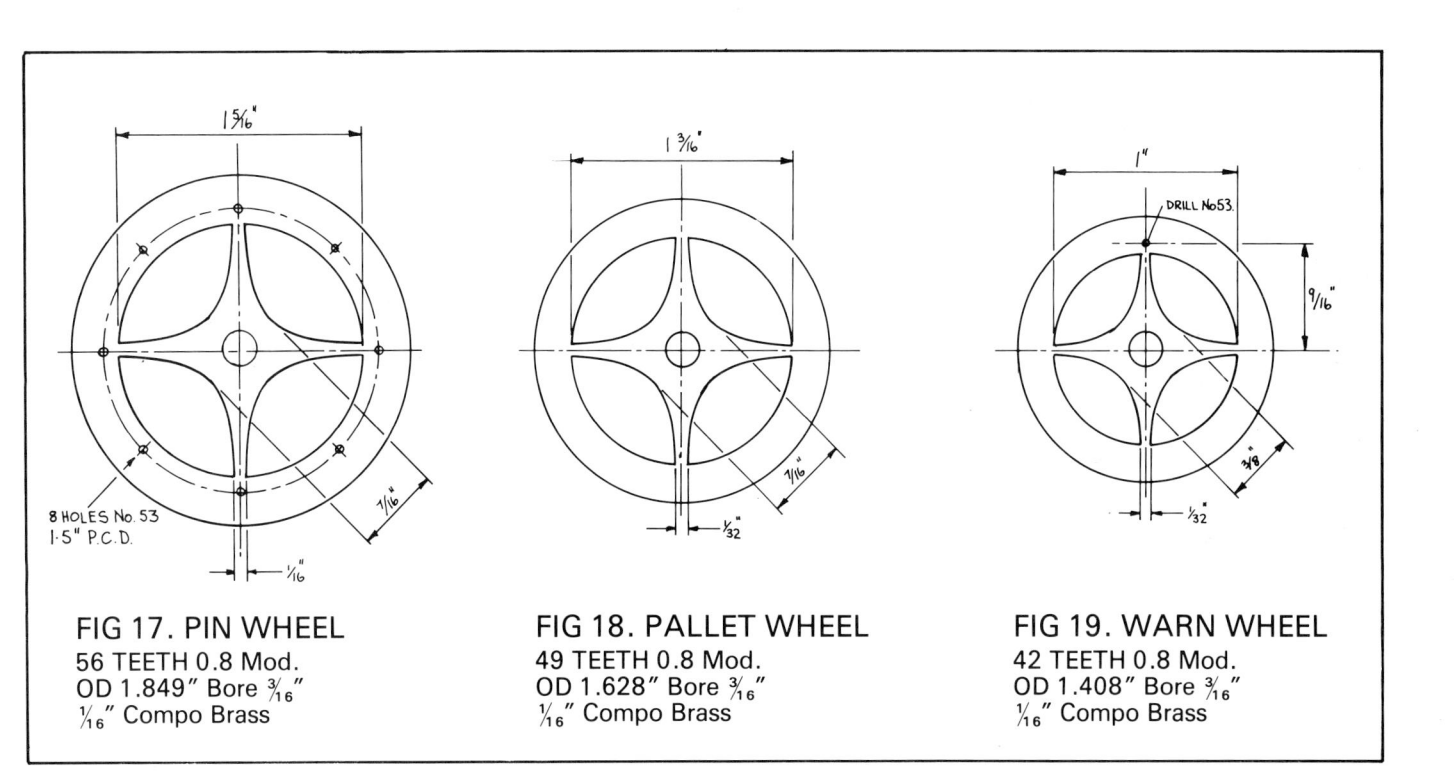

**FIG 17. PIN WHEEL**
56 TEETH 0.8 Mod.
OD 1.849" Bore ³⁄₁₆"
¹⁄₁₆" Compo Brass

**FIG 18. PALLET WHEEL**
49 TEETH 0.8 Mod.
OD 1.628" Bore ³⁄₁₆"
¹⁄₁₆" Compo Brass

**FIG 19. WARN WHEEL**
42 TEETH 0.8 Mod.
OD 1.408" Bore ³⁄₁₆"
¹⁄₁₆" Compo Brass

the rim. There is no hard and fast rule here, they just look right or not. Too thick looks clumsy, too thin looks weak. I find that about ¹⁄₁₆" is about right for the centre wheel, going a fraction thinner on the smaller wheels to keep the proportions correct. Hopefully, the drawings are printed full size so you will be able to trace the shape of the crossings and transfer them onto a piece of thin card or brass shim stock to make a template of one quarter of each wheel. Cut out the template and lay it on the wheel so that it lines up with the marks which define the width of the spokes and the 'base circle' already marked out. It is then an easy job to mark out each quarter on every wheel.

The waste metal is removed using a piercing saw with a fine blade. Drill a small hole in each corner where a spoke meets planes (sideways and fore/aft), as this will make cleaning up squarely much easier.

You will need a selection of files for crossing, and these are well worth buying. For this size of work, 3" or 4" crossing files of Nos. 4 and 6 cuts are very useful and I can recommend the Swiss made Grobet range. (See list of suppliers). An assortment of needle files is also useful. Knife edged and Barrette in particular are useful for finishing sharp corners. I stone down the backs of these files leaving a nice sharp edge to really get into the corners. Crossing files have two curved faces of different radii to file the rims and spokes. Use the face which fits the curvature best. It is also easiest to file out the sharp corners first, before starting on the curves. You will find that the smallest deviation from a true circle on the rim of the crossing will stand out like polished finish. Be very careful not to round off the tips of the teeth when polishing the wheel faces. Remember that the very best clockmaking tradition demands crisp sharp edges.

## WHEEL COLLETS

Four wheels altogether are fixed to their arbors by riveting to brass collets which are soft soldered to their arbors. The anchor is also mounted on a collet, so we need five in all. For convenience, all the collets are identical, as shown in the drawing, except for their centre holes. Three are drilled and reamed ⅛" and two are ³⁄₃₂". The remaining wheels are riveted to shoulders turned on the ends of their pinions. The collets are machined from ⅜" dia. brass rod. The only critical dimensions are a good fit on the arbors, a tight fit in the wheel holes and absolute concentricity of the wheel mounting spigot with the centre hole. The drawing shows the shape I make the 'decorative' part of my collets, but it was often a trademark of individual makers to make finely turned and decorated collets, so you may wish to design your own to make your clock that little bit different.

Hold the brass rod in a ⅜" collet and machine the ³⁄₁₆" long x ³⁄₁₆" dia. spigot using a tool ground to form the curved shoulder. You may find that the tool you made for the pillars will do. Take a light cut along the outer diameter of the rod to clean it up, then part off leaving enough to

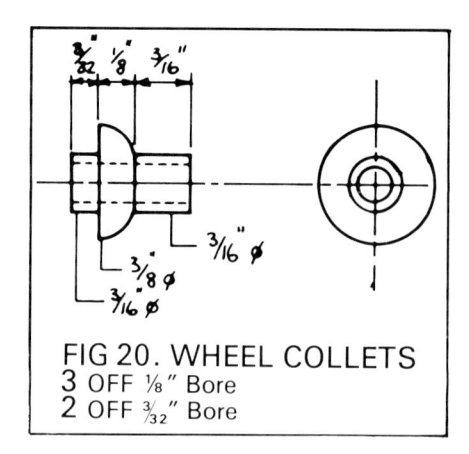

**FIG 20. WHEEL COLLETS**
3 OFF ⅛" Bore
2 OFF ³⁄₃₂" Bore

machine the opposite end. Repeat this process to make five collets. Next, holding them by the ³⁄₁₆" spigots in a lathe collet, turn the wheel seating, leaving the central boss ⅛" long. Slightly undercut the shoulders against which the wheels fit to ensure that the wheels seat dead square on their collets when they are riveted into position. Face off the ends of the collets to leave the wheel seats ³⁄₃₂" long, then centre drill, drill and ream the centre holes.

I find it convenient to put all the bits and pieces which start to accumulate into tins or boxes. I got fed up with making small pieces twice over after crawling round the workshop floor for a quarter of an hour looking for something which rolled off the bench!

# Chapter 5

## ARBORS AND PINIONS

Traditionally, arbors were made from pinion wire which was available from suppliers in a wide range of sizes and numbers of leaves. Unfortunately, it is no longer available, so we have to find another method. There are several ways of going about the problem. We can machine the arbor and its pinion from solid, fit a pinion onto a length of silver steel or make lantern pinions which consist of steel pins or trundles fitted into brass bobbins which are then fixed to silver steel arbors. I propose to discount the last method as it is completely out of context in a clock of this type. The first two methods produce exactly the same results visibly and obviously solid arbors and pinions are the nearest to the traditional parts. However, for the 'occasional' clockmaker, it is easier in the end to make separate pinion heads and fit these to silver steel arbors.

There are three reasons for this. Firstly, it is much easier to cut a length of pinion, drill a hole through its core, and cut it into the required lengths which can be either soft soldered or loctited to their arbors. Secondly, silver steel is easily available in the diameters we require. This saves turning fairly long slender shafts which can be tricky and it is also very easy to hold a length of silver steel in a collet to turn the pivots to size etc. The final and most important advantage is that it is much easier to 'depth' separate wheels and pinions if you haven't got a proper clockmakers depthing tool, since these are difficult to make and expensive to buy unless you intend making more than one clock. Simple depthing tools can be made easily and accurately and will ensure accurate depthing of wheels and pinions when we come to planting the trains between the plates. I will give complete details of a suitable depthing tool later.

Whichever method you choose, the first job is to produce the blanks for the pinions. As I have mentioned before, by far the best material is free cutting silver steel which machines easily and is also easy to harden and temper. All the pinions can be machined from $5/16''$ diameter stock. If you wish to make solid arbors you will need one piece for each arbor. Cut the stock over length for each arbor to allow for centre drilling each end. An alternative method here is to turn cones on each end of the stock which can then be supported with female centres during turning. I find it easiest to hold the rod in a $5/16''$ collet with the free end supported by a tailstock centre. I then turn down the long end of the arbor to diameter (see Plate 13). It is important here to turn to exact diameters given in the drawings as we have to hold the arbors in collets later to cut the pinions and machine the pivots etc. (The alternative method is to turn all the arbors between centres.) With the long end of the arbor turned to size, reverse in the collet and turn

*Plate 13. Turning the long end of the arbor.*

the opposite end to size, leaving the pinion head the correct length. Then holding by the long end of the arbor in a suitable collet, turn the pinion head to the diameter (see Plate 14).

As you will see from the drawings, some of the wheels are not mounted on collets but are riveted to the ends of pinions. In these cases, turn a shoulder on the end of the pinion head. This should be a tight fit in the centres of the wheel which is to be mounted on it.

The centre arbor is a special case. The front pivot is much larger than the rest and there is a long extension in front of the front plate to carry the minute wheel and pipe. The arbor between the plates usually tapers from about $3/16''$ diameter behind the front pivot, down to the root diameter of the pinion at the back end. I like to see this tapered arbor done properly, but I will give a much easier alternative later. To make things a little easier, it is convenient to make the centre arbor in two pieces as turning such a thin long shaft is tricky. The way to do it easily is to make the main part of the arbor, turned taper as in the drawing, with the front pivot drilled $3/32''$ diameter. A piece of $3/32''$ silver steel rod can then be silver soldered into the front of the arbor to form the front extension and save a lot of turning. If done well, the joint is invisible,

*Plate 14. Turning the short end of the arbor and pinion head.*

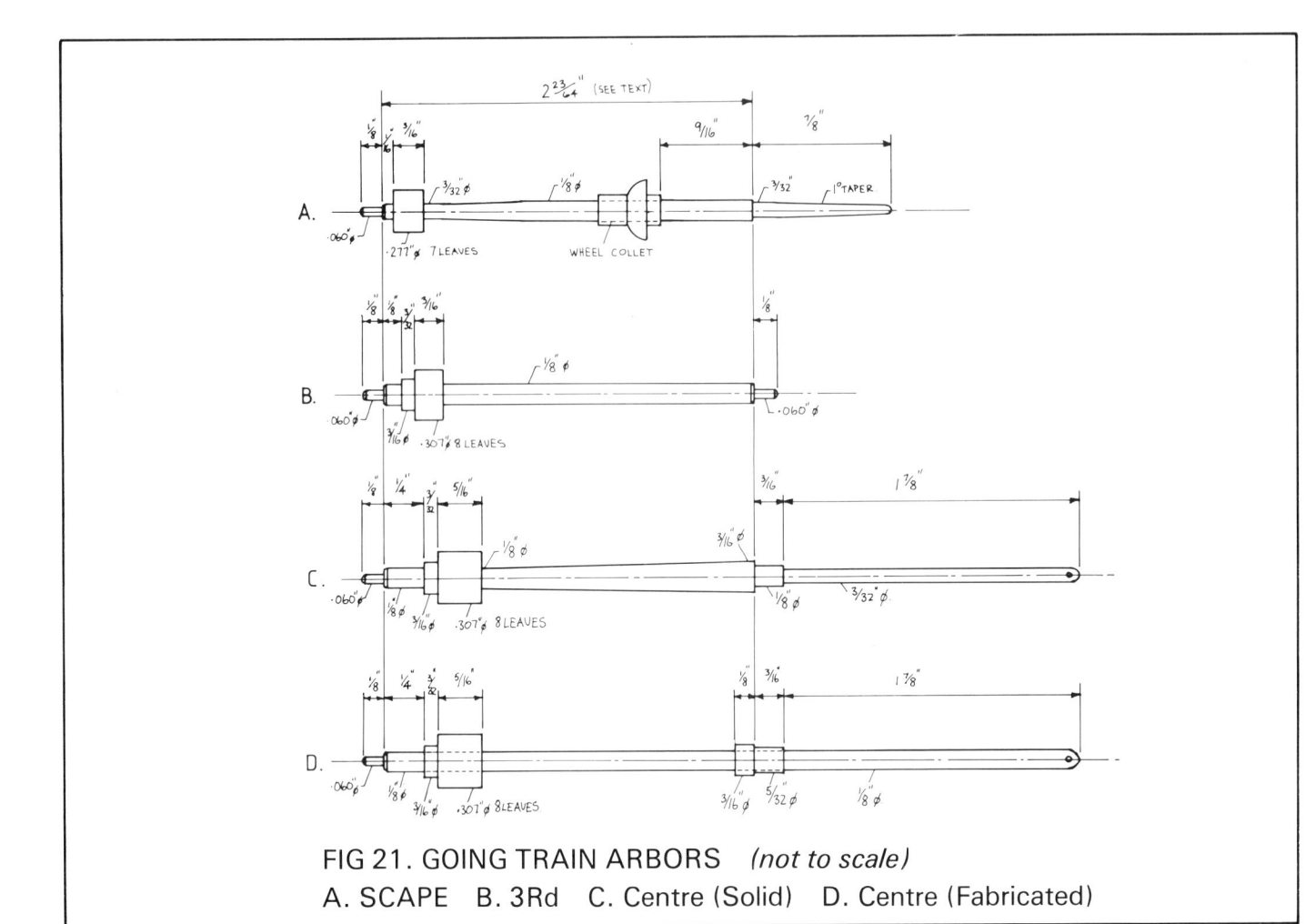

**FIG 21. GOING TRAIN ARBORS** *(not to scale)*
**A. SCAPE   B. 3Rd   C. Centre (Solid)   D. Centre (Fabricated)**

but you must be careful and ensure that the assembly is perfectly straight on completion.

With all the arbors turned to size, the pinions can be cut. As you will remember from the section on wheelcutting, pinions should be cut with a cutter which is *0.05 module smaller* than the wheels, but on a blank calculated using the *same module as the wheels*. Thus, in the drawings, the dimensions given are for 0.8 module pinions, but we cut the leaves with a 0.75 module cutter. As I mentioned before, this gives a more traditional leaf form and produces leaves which are thicker at the roots and are therefore stronger.

Plate 15 shows my set-up for cutting the pinions on the horizontal mill using a dividing head. The arbor is held in a collet with the free end supported by the tailstock centre. A similar set up can be used in the lathe, holding the arbor in a collet in the headstock, with the headstock dividing attachment fitted, and the cutter held in a spindle mounted on the vertical slide.

For cutting pinions, it is vital that the cutter is set absolutely central to the blank. If the cutter is slightly off centre, the pinion leaves will lean to one side, so it is worth taking some trouble to make sure that the set-up is accurate. Set up the drive to give about 150 to 200 rpm. If you are cutting pinions on the lathe, a large pulley can be made from plywood and fitted to the spindle, using a small pulley on the motor to give the correct speed. I usually take three cuts to form the leaves, using a slow feed and plenty of coolant. Two roughing cuts and a fine finishing cut of two or three thous finally form the tips of the leaves. If you are flycutting the pinions using the cutters described earlier, it is probably best to gash cut the leaves using a slitting saw to remove most of the waste and then finish off with two or three passes with the flycutter to form the leaf shape. Whichever cutter you use, the finishing cut should be taken with a very slow feed to give as good a finish as possible from the machine to save work later polishing the pinions.

With the leaves cut, saw off the centre

*Plate 15. Cutting pinion leaves.*

holes on the ends of the arbors and return to the lathe for turning the pivots. Plate 16 shows a pivot being turned. The long end of the arbor is put into the back of the Myford collet and pushed through so that just enough is projecting to turn the pivot. As with the barrel arbors, turn the pivot leaving slightly oversize to allow for polishing and burnishing with the pivot file/burnisher. The ends of the pivots should be rounded neatly, the pivots should be

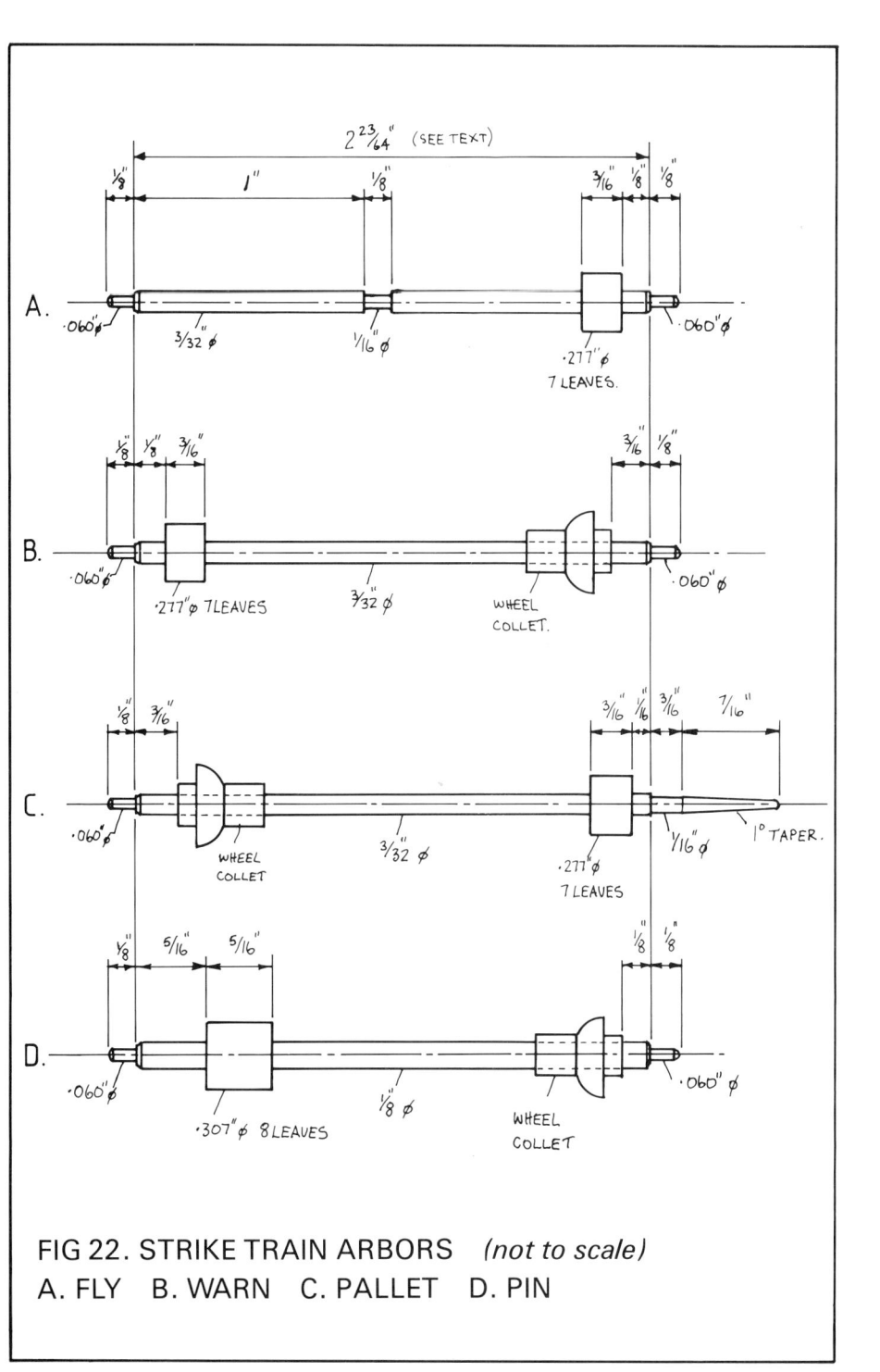

**FIG 22. STRIKE TRAIN ARBORS** *(not to scale)*
A. FLY   B. WARN   C. PALLET   D. PIN

parallel and have a high polish with no marks. The ends of the arbors should be tapered slightly to a clean crisp shoulder. When one pivot is finished, reverse the arbor in the collet and do the other end with the long end of the arbor in the collet and the pinion pushed hard against the face of the collet. Take very light cuts to avoid bending the arbor and finish as just described. Check the shoulder length by laying the arbor across the plates. There should be about $\frac{1}{64}$" end float. It is best to err on the tight side and adjust as necessary after trying the fit on your plates.

As I said earlier, solid arbors and pinions demand the use of a proper depthing tool as the whole arbor must be meshed with its adjacent arbor to determine the pivot centres on the plates. It is much easier to make separate pinion heads and solder or Loctite these to silver steel arbors later, as you can then mesh the pinion head and its wheel in a very simple accurate tool which you can make yourself. All the arbors are $\frac{1}{8}$" or $\frac{3}{32}$" diameter. The 8 leaf pinions will easily accept a $\frac{1}{8}$" reamed hole through their core and the 7 leaf pinions will just take $\frac{3}{32}$" hole with my cutters. To make pinion heads, turn a length of steel to diameter and cut a length of pinion on its end, using a set up as already described. As you are holding a much larger diameter rod, the set-up is more rigid. Cut enough pinion to make two or three pinion heads and then return the job to the lathe. Drill the core through, the required diameter, and ream, before carefully parting off the pinion heads. Use a thin parting tool with a slow feed to avoid bending the leaves. Burrs can be removed with a knife edged needle file.

With this method, the arbors are made from lengths of $\frac{1}{8}$" or $\frac{3}{32}$" diameter silver steel with the pivots turned on the ends as already described. After marking out the pivot holes from the depthing tool (to be described) the pinion heads can be either soft soldered or loctited to their arbors. Plate 17 shows a finished solid arbor along with two pinion heads and a silver steel arbor. With this method, the centre arbor can be made of one length of $\frac{1}{8}$" diameter silver steel, with a pinion head fixed to the back end and a collar fitted as shown in the drawing to form the front pivot. Again, this is a much easier method which works perfectly well, but it doesn't look as nice as the traditional tapered centre arbor.

Pivots and pinions should be hardened and tempered to prevent wear. Surprisingly, on every antique clock I have ever restored, the brass wheel teeth usually show very little sign of wear, but very often the pinions are very badly grooved where they mesh with the wheels. This seems odd at first because the pinions are much harder than the wheels. The reason for this wear is that over the years, dust embeds itself into the soft brass wheel teeth. The dust is abrasive and thus turns the wheel into a lap which gradually grinds away at the hard steel pinion leaves. This is one reason why wheel teeth should not be oiled as oil would encourage the dust to stick to the wheels and worsen the problem.

Solid arbors can cause problems with distortion during heat treatment. The best way to overcome this problem is described in De Carles book *Practical Clock Repairing*. The whole arbor and pinion is bound completely with soft iron binding wire, using plenty of wire. This is then rubbed all over with a piece of soap to fill all the grooves formed by the wire and these exclude air from the arbor. The whole parcel is then heated to cherry red as evenly as possible, before quenching LENGTHWISE (ie, vertically) into water. Quenching vertically minimises distortion, as the arbor is cooled evenly around its circumference. Unwrap the arbor, dry it off and test with a fine file to make sure it is hard. If it is, bind it up in wire again and dip into some thick oil. Wave the arbor in a gentle flame until the oil vaporises and black smoke is given off before quenching again. It sometimes helps to float a layer of oil about $\frac{1}{4}$" thick on top of the water used for quenching. (I use an old bean can with the lid cut off for quenching.) The oil on top of the water tends to take the shock out of quenching and avoids undue stresses which cause distortion. Arbors that do bend can be straightened by gently tapping them true on a steel block. They can be checked for truth by holding in a collet in the lathe and spinning the arbor.

Once again, separate pinions and arbors are easier to harden without distortion. We only need to harden the pinion head and the pivots on the ends of the arbor and this can be done by normal methods with very little risk of distortion.

The final stage on the arbors is to polish them up and polish the pinion leaves to remove tool marks. The arbors are easily done with fine emery sticks and the pivots should be burnished to a high polish. The pinion leaves are polished using a piece of hardwood cut to a vee like the end of a chisel. The pinion is held on a block of wood with a groove cut in it. Hold the block in the vice, rest the pinion in the groove and charge the end of the polishing

*Plate 16. Turning pivots.*

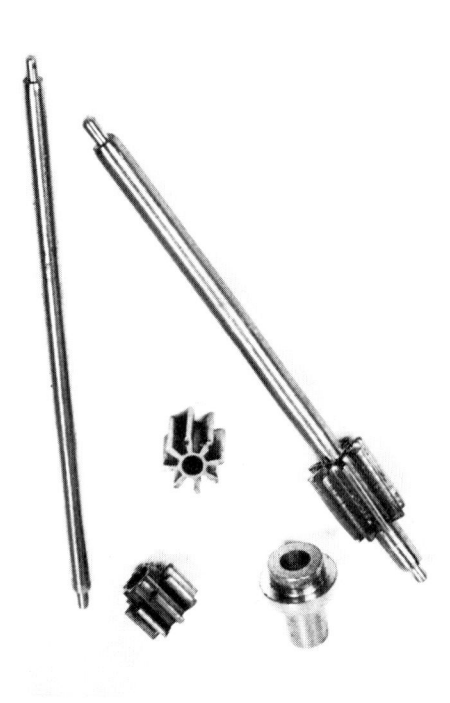

*Plate 17. Solid arbor, pinion heads and wheel collet.*

stick with fine abrasive paste and oil. (I use fine valve lapping compound.) Rub the stick backwards and forwards in the leaf spaces and it will soon take up the shape of the pinion leaves. Polish all the marks out of the leaves, re-cut the end of the stick and repeat, using fine Diamantine powder or rouge powder to bring it to a high polish.

The aim is to polish the leaves without altering their form or rounding off the ends of the leaves. The job doesn't take long and is worth doing well as it makes the meshing smoother and looks good.

One final point — don't fit the pinion heads or wheels onto the arbors yet if you have made separate pinions — we need the pinions and wheels off the arbors for depthing which I will describe next.

**Important:** Some makes of pinion cutters will cut deeper than others — check before drilling and, if necessary, reduce the diameter of the arbor to suit.

*Plate 18. A group of finished arbors and wheels.*

# Chapter 6

## DEPTHING THE WHEEL TRAINS

I have decided to describe the depthing operation before making the escapement, because we need to use a depthing tool to mark out the shape of the pallets. So if I cover all the points about depthing first, you can make your own decisions about which type of depthing tool to make or buy.

Plate 19 shows a traditional type of Depthing Tool produced by J. M. Wild of Sheffield. These are available with either aluminium or brass bodies and are beautifully well made. Mr Wild also supplies castings for a tool which he described in the *Model Engineer* during 1973. Part machined kits are also available from other suppliers.

The traditional type of tool is undoubtedly the best for the job and is the only method of depthing traditional solid arbors. As you can see in the photograph, the tool consists of two pairs of 'runners' which are pointed at one end for scribing arcs on the clock plates. The runners also have hollow centres formed on their inner ends so that arbors can be held between centres. The runners are carried in two rigid castings and can be adjusted to accommodate arbors of different length. The castings are hinged at the bottom and are spring loaded against an adjusting screw.

In use, an adjacent pair of arbors and wheels are mounted in the runners as in plate 20. They are adjusted endways so that the wheel lines up with the pinion with which it will finally mesh in the clock, and so that the arbors spin freely without shake. The wheel and pinion are then brought into mesh using the adjusting screw until the correct depth is obtained.

The correct depth is very difficult to put into words! The method I use is to bring the wheel and pinion into mesh until they mesh too deeply. This can be felt very easily by revolving one arbor when it will be found that the mesh feels tight and 'lumpy'. Open the tool slowly whilst revolving the meshing wheel until the action feels smooth. It helps if you apply a little pressure to the pinion arbor with a finger while you check the mesh. The aim is to get the wheel and pinion to mesh as deeply as possible, consistent with smooth running, ie, the wheel and pinion should mesh on their pitch circle diameters. Examine the action using an eyeglass, looking at the end of the pinion. Excessive backlash means that the mesh is not deep enough. A few minutes experimenting with a pair of arbors in the tool will soon put you right. As I mentioned earlier, it is very difficult to define feel, but you will know when things are correct when you try it yourself.

With the tool adjusted for correct depth, it is now a simple matter to mark the arbor centres onto the plates prior to centre punching and drilling. The only point you need to watch when marking out is that the runners are adjusted so that the tool is exactly upright when scribing the arcs, particularly when you are using the cone ended runner to mark from a hole already drilled. It is very easy to get a false measurement unless the tool is exactly square to the plates. It is also wise to plant one arbor at a time, and check the mesh with the plates assembled each time to avoid any accumulation of mistakes as we go along.

If you opted to make separate pinion heads, the depthing tool problem is much easier to solve. As the wheels and pinions all have holes through their centres, we can make a very simple but effective tool as shown in the drawing, Fig 23, and the photograph, Plate 21. This tool also comes in handy for marking out and testing the escapement, so it is worth making anyway. The tool is similar to one illustrated in Britten's *Watch and Clockmakers Handbook*.

The tool consists of two bushes mounted on a brass plate, one fixed and one ad-

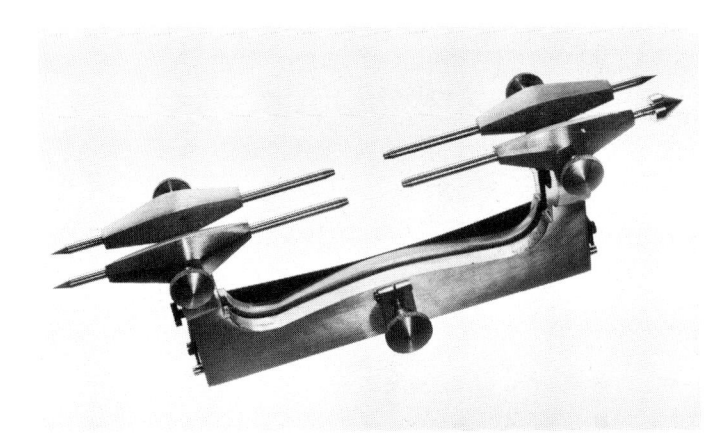

*Plate 19. The J. M. Wild depthing tool.*

*Plate 20. A pair of arbors set up in a Colin Walton depthing tool.*

*Plate 21. A simple home-made depthing tool.*

**FIG 23. A SIMPLE DEPTHING TOOL**

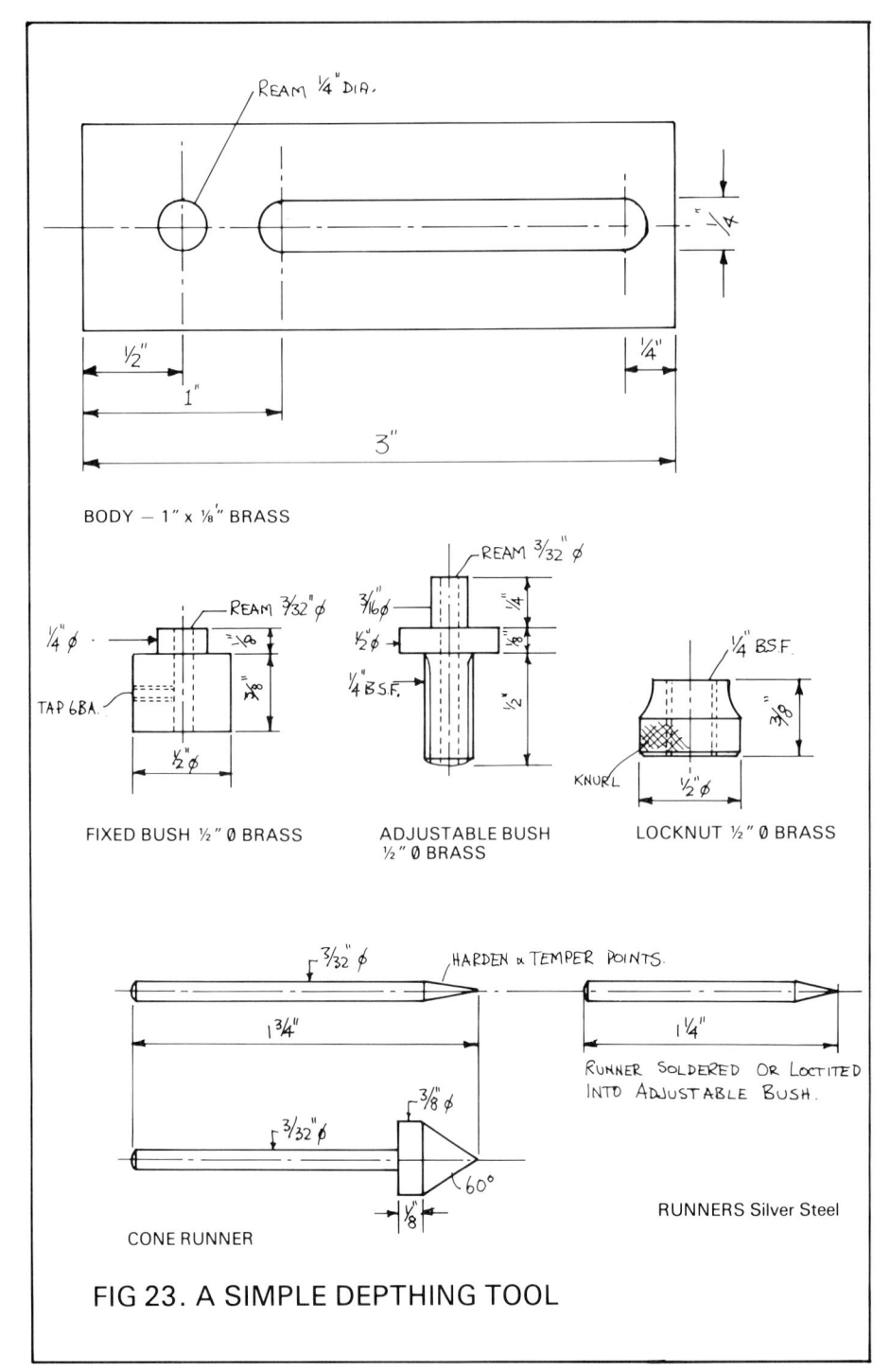

justable along a slot in the plate. These bushes carry $3/32''$ diameter silver steel runners with hardened and tempered points, or a runner with a cone point for locating in holes already drilled. A pointed runner is fixed permanently in position in the adjustable bush, and the fixed bush is fitted with a lock screw so that the cone runner can be inserted when required and the points of the runners can be adjusted to ensure that the tool is square to the plates when marking out. The construction of the tool is obvious from the illustrations. The point to watch is that the spigots and holes for the runners in the bushes are concentric and square to each other. Make sure that the fixed bush seats squarely on the plate and make sure also that the silver steel used for the runners is straight. The cone runner is best made by soldering a piece of steel onto the end of a piece of $3/32''$ diameter silver steel, and then turning the cone with the runner held in a $3/32''$ collet to ensure concentricity.

In use, the pinion is fitted onto the runner projecting from the top of the fixed bush and the wheel is placed onto the spigot on the adjustable bush. The great wheels and hour wheel have larger centre holes, so you will need to turn up a couple of bushes to fit the wheel holes tightly, with $3/16''$ holes through them to fit the spigot on the tool. The mesh of the wheel and pinion is then adjusted by moving the adjustable bush along its slot. Fine adjustment can be achieved by tightening the locknut finger tight, and tapping the bush along its slot from underneath. If anything, this tool is more accurate than the traditional pattern, as it is easier to make accurately and it is more rigid and therefore less likely to flex. It is obviously impossible however to use this tool to depth arbors with solid pinions.

Fig 24 shows the layout of arbors on the frontplate. The sequence of planting the trains that I use is as follows:

## GOING TRAIN

Fit the two plates together, locating with the register pins already fitted, and decide which is to be the front plate. I usually choose the plate with the most scratches as the front one since this is behind the dial and is not seen, so any bad marks are not quite so important. Mark an accurate centre line up the long dimension of the front plate. All marking out of the depthing will now be done on this plate.

Start by marking the position of the centre arbor, $2¾''$ up from the bottom of the plate on the centre line. Make a very fine centre punch from a piece of $1/8''$ diameter silver steel and centre punch this position. Next, mark out the position of the 'scape wheel arbor which is again on the centre line, $2 1/16''$ up from the centre arbor. Again, lightly centre punch this location. Scribe a light line across the plate $2¼''$ up from the bottom, at right angles to the centre line. This is the line on which the barrel arbors will be situated.

Set up the centre pinion and the going train great wheel in the depthing tool. If you are using a traditional depth tool, you will have to make a false arbor to mount the great wheel. This consists simply of about $1''$ of bar, turned to be a tight fit in the great wheel holes with a 30° cone centre turned on each end to fit into the female centres in the depth tool runners. Adjust the depth for correct mesh as already described then, with one point of the tool located in the centre arbor punch mark, adjust the runners so that the tool is square to the plate.

Scribe an arc to cut the line scribed across the plate to the right side of the centre line. Where this arc crosses the scribed line is the position of the barrel arbor for the going train, and this point should be carefully centre punched. As we want the two barrel arbors to be equally spaced about the centre line of the clock, the strike train barrel arbor hole can also be marked. It is positioned as shown, on the horizontal line. On my clock, these two arbors are $1 9/16''$ either side of the centre line, but this dimension depends on the depthing of your wheels so check this measurement from the job.

With the plates still pinned together, transfer them to the drilling machine. The barrel holes are drilled and reamed $¼''$ diameter. Start with a small drill and open up the holes gradually to reaming size. Make sure that the reamer goes through exactly square to the surface of the plates. Separate the plates to remove any burrs that have formed on the inside. A couple of twists with a countersink should do the job. Try each barrel arbor pivot in its hole, and if you polished the pivots as described earlier, they should all be a nice running fit. If all is well, fit the plates together again to drill the centre pivot holes, making sure that the plates fit exactly together with no dirt or swarf trapped between them. Check the finished size of the back pivot of your centre arbor and select a drill about 0.005″ smaller to drill the hole. If your pivot is 0.050″, a No 56 drill will be about right. This leaves a few thou to open out with a broach to fit the pivot exactly. Drill through both plates with this drill then lift off the front plate and open up the centre hole to about 0.005″ smaller than the front pivot diameter.

The holes are opened out to size with five sided cutting broaches which are obtain-

*Plate 22. Using the simple depthing tool.*

able in a range of sizes, (see list of suppliers). Each broach is tapered and is used just like a reamer to finish the holes to size. Keep taking a small cut and trying the pivot in the hole between cuts until the pivot will enter about half way through the thickness of the plate from the inside surface. Then broach out the hole from the other side of the plate until the pivot fits.

Round, tapered polishing broaches are available which burnish and harden the walls of the holes, thus improving their wear characteristics. If you intend to use a polishing broach, open out the holes until the pivots enter but are tight and then burnish the holes out to size.

The correct fit is again difficult to define. Model engineers seem to make pivots too good a fit! On several occasions I have had visits from people who have made a clock and produced a beautiful piece of work that just didn't want to run. The problem is invariably that the pivot holes are just that little bit too tight and a quick broach out of the holes soon saves the day! Of course, the holes must not be too loose — there is no point in building in instant wear!

A good guide to the correct fit is that the pivots should fit the holes without perceptible play, but they should just rock from side to side slightly. Another guide is that when the arbors are assembled between the plates they should revolve freely and silently. If you give them a good spin with your fingers they should gradually slow down and finally stop. If they stop suddenly or quickly, they are probably too tight or there is dirt in the holes, so check each arbor on its own and get it right before starting the next. A final guide to the fit of

*Plate 23. Split stake and punches.*

pivots is that if you quickly invert the plates with the arbors in position you should hear an audible click as the arbors drop, taking up their endshake.

With the first pair of holes in the time train drilled, you can try the great wheel/barrell assembly and centre arbor between the plates. If you have made separate pinion heads, soft solder the pinion onto the centre arbor in the position shown in Fig 21. To fit the wheels onto their seats you will need to make a split riveting stake and a couple of hollow punches. The stake is easily made from two pieces of mild steel about ½" square and 4" long. Fit a ¼" steel dowel into each end of one bar to locate in clearance holes in the other bar so that the two can be assembled accurately. Hold the two bars together in the drilling machine vice and drill a series of holes along the joint line between the two bars. My stake has a range of holes from $\frac{3}{32}$" to ½" diameter. The finished stake should have semi-circular holes along each half of the joint face as in the photograph. The photo also shows a couple of punches made from ¼" diameter silver steel. I make these with all sorts of ends to suit the job. For riveting wheels mounted on ⅛" arbors, drill the punches about $\frac{5}{32}$". Make one punch with a flat end and one with a convex end as in the drawing, Fig 25.

To rivet a wheel onto a pinion (as on the centre arbor), mount the arbor in the stake so that the pinion head rests on top of the stake. Select the smallest hole in the stake that will accept the arbor without gripping it when the stake is held in the vice. Place

**FIG 24. LAYOUT OF ARBORS ON FRONT PLATE**

33

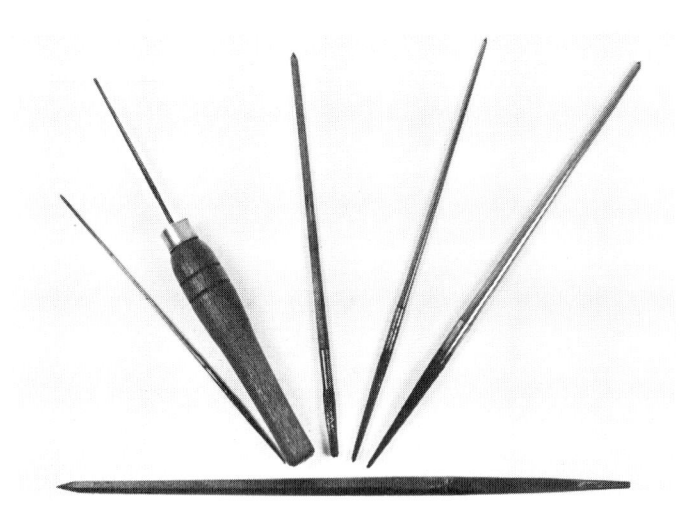

*Plate 24. A selection of broaches.*

the wheel in position over its seat and tap it down onto the shoulder. Remove the arbor from the stake and spin it between the fingers to make sure that the wheel spins true without wobble. If it does, replace it in the stake and using the convex ended punch, gently peen over the end of the pinion which should be just above the surface of the wheel. Revolve the arbor in the stake while you rivet to make sure that you

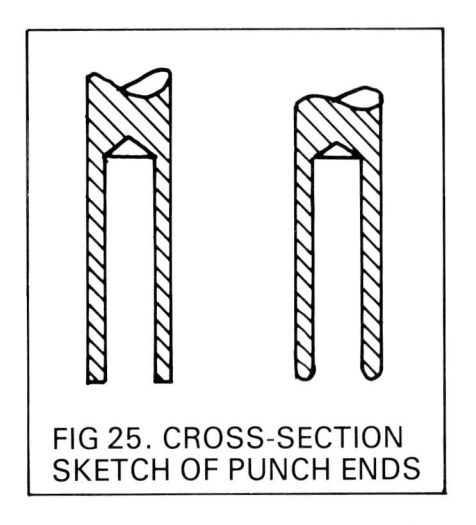

**FIG 25. CROSS-SECTION SKETCH OF PUNCH ENDS**

do it evenly. Also check to make sure that the wheel doesn't lift off the shoulder as you are riveting. Finish off with the flat ended punch which will spread any metal raised during the first stage and leave a neat finish. Always use the lightest possible hammer blows. Traditionally, brass wheel collets were mounted on their arbors before the wheel seats were turned, but with accurate reamers I find it easier, and just as good, to rivet the wheels to their collets before soldering the collets onto the arbors. If you run a reamer through the collet hole after riveting and the arbors are turned accurately you will have no problems with eccentric wheels.

If you are using the simple depth tool, make sure that you have depthed each wheel BEFORE mounting it on its arbor. Of course, if you are using the traditional depthing tool, you must assemble the arbors and wheels before you do the depthing.

When I make a movement with pillars riveted into the backplate, I make up a set of 'trial' pillars. These are simply lengths of ½" diameter brass rod with spigots turned on their ends to fit the plates. The spigots are threaded so that the plates can be temporarily assembled to check the depthing as I go along. These trial pillars must, of course, be exactly the same shoulder length as the actual pillars that you will finally use.

When you have planted the barrels and centre arbor correctly, drill the escape arbor holes in exactly the same way as you did the centre arbor holes. Again, check the 'scape arbor between the plates before the next stage. The third wheel arbor must now be depthed from both the centre and 'scape holes. Set up the centre wheel to mesh with the third pinion in the depth tool and adjust for correct mesh. Using the cone runner in the centre hole, scribe an arc just to the right of the centre line. Then remove the centre wheel and the third pinion from the tool and replace them with the third wheel and 'scape pinion. Adjust for correct mesh and scribe an arc from the 'scape hole to cross the first arc. Where the two arcs cross is the position of the third arbor. Centre punch this position and drill as before. You can now try the complete time train between the plates, minus the 'scape wheel and pallets which we have yet to make.

Depthing the strike train is straightforward. The pin wheel and pallet arbors are situated on a line 1½" in from the left hand edge of the plate as in Fig 24. The warn wheel is 1 7/16" in and the fly is 1 1/8" in from the same edge. Mark these lines lightly on the plate then, working up from the great wheel/pin wheel pinion, depth each pair of arbors in turn, locating each arbor where the arcs scribed with the depth tool cross the lines marked on the plates. Don't forget to check each pair of arbors as you go along.

As a final word on depthing, this is probably the most important aspect of clockmaking and poor depthing probably causes more trouble than anything else. It pays, therefore, to be particularly careful at every stage to ensure accuracy. It is far easier to take time getting it right first time than it is to fill wrongly drilled holes and reposition them!

*Plate 25. The time train of the finished clock.*

# Chapter 7

## MAKING THE ESCAPEMENT

This clock has the traditional Recoil or Anchor escapement, which to this day is still the most popular escapement for domestic clocks. This escapement was invented in the latter part of the seventeenth century by Dr Hooke. Its invention brought about the use of the Royal or seconds pendulum, and brought about a vast improvement of the accuracy of clocks, which up until then had used the Verge escapement, and often varied by as much as fifteen minutes in a day! The advantages of the recoil escapement for our purposes are that it has a good performance, (even if not quite correctly made), it is the easiest of all the escapements to set out and make, and it is the easiest to set up and maintain. It will also work after many years of use have produced considerable wear. Indeed, I often get clocks in for restoration with very badly grooved pallets, which still run and keep reasonable time — I doubt if some of these clocks have ever been fully serviced since they were made 200 years or so ago.

The 'scape wheel has 30 teeth, and is shown in Fig 26. The wheel teeth are cut on a blank 1.65" diameter. As you will see

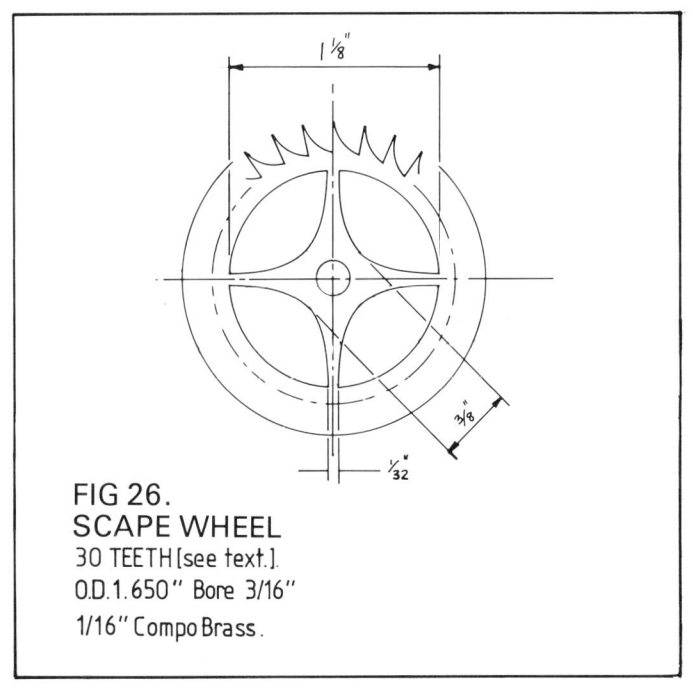

**FIG 26.**
**SCAPE WHEEL**
30 TEETH [see text].
O.D. 1.650" Bore 3/16"
1/16" Compo Brass.

from the drawing, the backs of the teeth are radial and the fronts curve down from the tip to meet the root of the radial face. On antique clocks, the radius of the curved face of the teeth varies considerably from clock to clock, but it is now usual to make this radius the same as the radius of the escape wheel itself. The tooth depth should be about the same as the distance between the tips of two adjacent teeth. I cut the teeth with a No. 6 recoil 'scape wheel cutter, as supplied by Richards of Burton, but you can easily make a flycutter from 1/4" diameter HSS or silver steel. These cutters are illustrated in Plates I and II of the gear-cutting appendix

To cut the wheel, set up as with the other wheels, making sure that the tip of the cutter is exactly central to the blank. Cut two teeth a little on the shallow side, and gradually cut both teeth deeper until there is only a slight witness left on the tip between the two cuts. On no account must you leave a feather edge as this would leave the tips too weak. About 0.010" is a reasonable tip thickness. When you have got the set up right, go ahead and cut all 30 teeth.

To mark out the Pallets we need to mount the 'scape wheel and the Pallet material in the simple depthing tool described earlier. In this way we can mark out the shape of the pallets directly onto the material — I think that this method is much more accurate than drawing the pallets on paper and sticking this onto the pallet blank. It may be as well, however, for the newcomer to clockmaking to draw the pallets on paper first, in order to familiarise himself with the process, before starting in metal.

The illustrations and photographs show the various stages involved in marking out the pallets. I advise you to study these carefully as the job progresses to avoid mistakes.

Set up the depth tool so that the two runners are exactly 1 5/32" apart, centre to centre. If you have drilled the holes in the plates for the 'scape and pallet arbors, this operation can be done easily by setting the runners to the holes — otherwise you can set the runners very accurately by engaging the points of the runners in the engraved markings on a good steel rule. At all costs, it is vital that the runners are exactly the same distance apart as the holes in the plates or that the holes are drilled after marking out their centres using the depth tool set to the correct measurement. The pallets are made from a piece of high carbon steel, 1/8" thick. I use precision ground flat stock, but an excellent alternative is an old file which can be annealed and filed flat before starting. You need a piece about 2" × 1 1/4". Coat the blank with layout blue and draw a centre line (A-B) across the short face of the blank. About 3/8" down from the top, on the centre line, centre punch and drill 3/32".

The blank is now ready to mount on the depth tool for marking out. It will help later if the top of the fixed runner is centre punched exactly centrally to locate the point of a pair of dividers. As you will see in the photograph, the pallet blank is placed on the fixed runner, and is clamped to the tool using a small pair of toolmakers clamps, so that the centre line drawn on the blank would pass exactly through the centre of the other runner if it were produced. The 'scape wheel is then placed on the spigot of the adjustable runner. You may need to put washers under the wheel or the blank so that the wheel rests on the

*Plate 26. Marking out the pallets.*

blank.

The method used to mark out the pallets is the one described in Gazeley's book 'Watch & Clock Making & Repairing'. Revolve the wheel until the radial face of any tooth is exactly on the centre line on the blank and hold it firmly in position. Count round four teeth either side of the centre and mark a dot at the tips of these two teeth, (points C & D). Revolve the wheel anticlockwise exactly half a tooth space (ie, the centre line now passes half way between two teeth), and again mark dots on the blank at the tips of the fourth tooth either

**FIG 27. A – E MARKING OUT THE PALLETS**

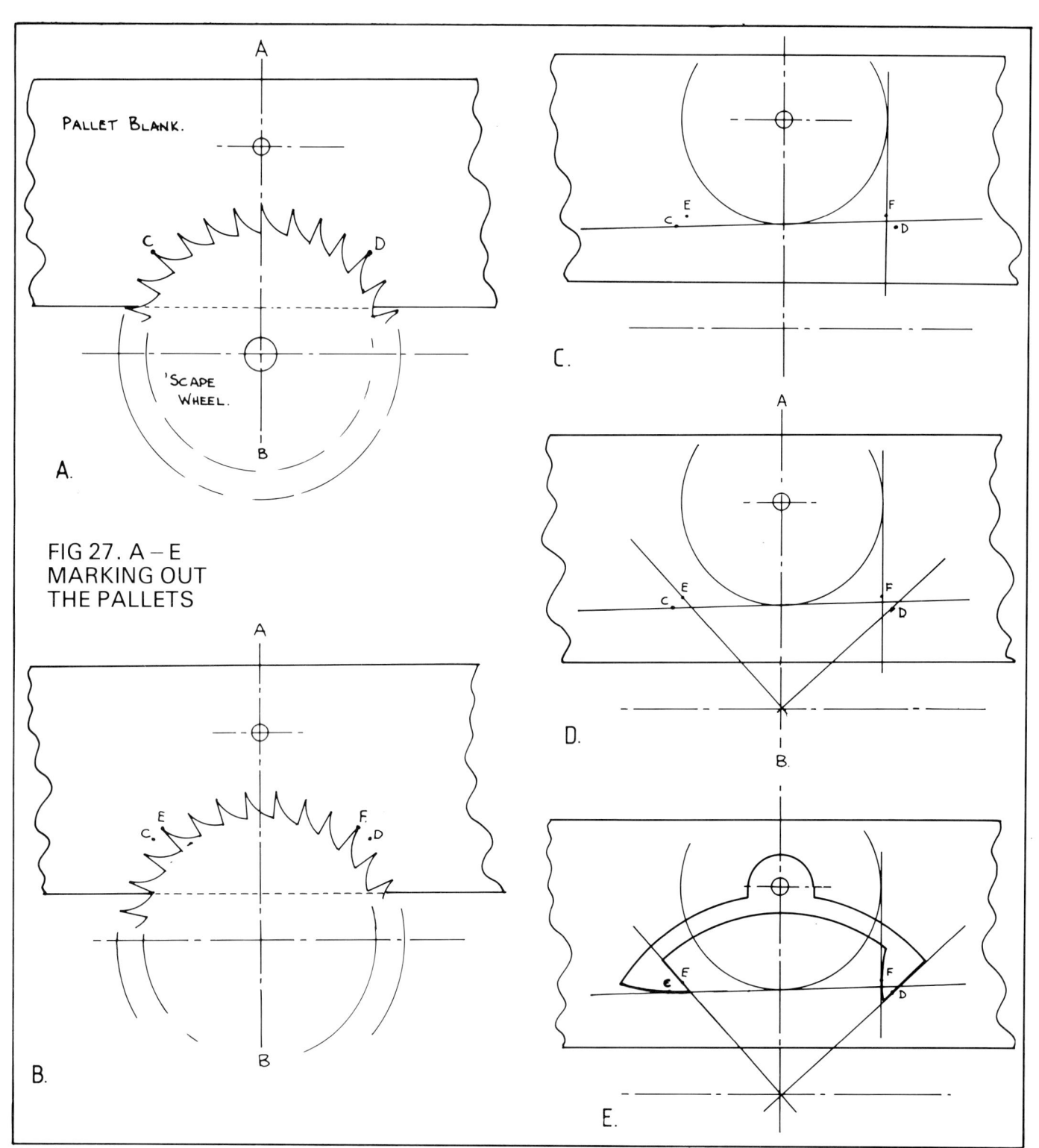

side of the centre line, (points E & F). These four dots mark the points of engagement and dropping off for the entrance and exit pallets.

To obtain the shape of the nibs of the pallets, proceed as follows. Remove the wheel from the depth tool, leaving the blank in position. Scribe a line passing through point C and half way between points F & D. This line gives us the impulse face of the entrance pallet. Next, using dividers located in the centre dot on the end of the fixed runner, scribe a circle from the pallet hole centre to just touch the line just drawn. Now scribe a tangent to this circle passing through point F. This line will give us the impulse face of the exit pallet. The backs of the nibs are marked by scribing radial lines from the centre of the 'scape wheel passing through points E & D.

The arms of the anchor are parts of a circle whose centre is at the centre of the 'scape wheel. Scribe the inner arc first, allowing clearance for the 'scape wheel teeth, then scribe the outer arc leaving the arms of the anchor about $\frac{3}{32}"$ thick. The boss round the pallet hole should be just over $\frac{3}{8}"$ diameter, as the collet with which we shall mount the anchor onto its arbor is $\frac{3}{8}"$ (see Fig 20). The mounting hole in the anchor will need to be opened out to $\frac{3}{16}"$ to fit the collet, but it is convenient to leave this until later as we can test the action of the escapement in the depth tool before fitting the collets, etc.

With all the marking out done, carefully cut out the anchor. You will probably find it easiest to chain drill round the inside and saw the outside, keeping away from the lines to allow for cleaning up. With the shape roughed out, file to the lines VERY CAREFULLY, making sure that all edges are kept perfectly square, especially the acting faces of the nibs. File the acting faces of the nibs to a slight curve as indicated on the drawing.

When this is done, fit the 'scape wheel and pallets back into the depth tool so that they mesh together, using packing as necessary, and try the action. At this time, you will no doubt be disappointed that the whole thing is tight everywhere and just doesn't work. Don't throw it all into the scrap box, yet! All that is needed is a little very careful filing. We do not want to alter the acting faces as they should be correct as marked out. Take the pallets out of the tool and file a little off the backs of each nib, i.e. the faces on which points E & D are situated. Do this a little at a time, removing equal amounts from each side. Try the action again, and then remove a little more — again equally — until the teeth of the 'scape wheel will just pass the tips of the nibs when the anchor is rocked from side to side. If too much metal is removed, the teeth will drop too much when the clock is running and this will mean loss of impulse.

When the pallets are just free, harden the nibs. I do this by holding the boss at the centre of the anchor in a pair of pliers and hold the anchor over a flame from the blowpipe, so that the flame heats both nibs at once. Heat to red hot and quench in brine. Leave the nibs in the dead hard state — there is no need to temper them. If you now polish the acting faces of the pallets with very fine emery sticks until they are super fine finished, there will be enough clearance for the escapement to work. The remainder of the anchor is cleaned up with emery sticks and by rubbing its faces on a sheet of emery cloth until all marks are removed. A straight grained finish on steel parts looks well, or you can go the whole hog, and polish all over. Take special care, however, to keep all corners sharp and all edges square — especially the acting faces. If these are not square, the action of the escapement will be affected by endshake in the pallet arbor.

When you have finished, if you mount the escapement back in the depth tool and hold the tool vertically, the anchor should rock back and forth when you revolve the 'scape wheel clockwise.

*Plate 27. The 'scape wheel arbor, pallets, crutch and backcock, ready for assembly.*

$\frac{1}{16}$" diameter steel rod, silver soldered to the shaped plate which will eventually accept the brass block on the pendulum. This slotted plate is made from $\frac{1}{16}$" mild steel. The crutch is a straightforward sawing and filing job — the slot being filed to shape after chaindrilling the waste out. Stamped crutch blanks with the slot formed and the rod attached are available however from the suppliers already listed. The final operation on the crutch-arbor assembly is to soft solder the crutch rod into the $\frac{1}{16}$" hole drilled in the square boss on the arbor. Make sure that the crutch plate is square to the arbor, so that when the plate is bent up, it is parallel to the arbor in all planes. The photograph shows the finished assembly with the anchor fitted. Do not fit the anchor until the back cock has been fitted to the backplate, as we need to fit the arbor through a $\frac{1}{8}$" hole in the backplate in order to position the back cock.

### THE BACK COCK
The back cock, which is made of brass, is shown in Fig 29. Traditionally, these were castings, and some suppliers stock rough castings for the job, but I usually make my own up. The easiest way to do this is to fabricate the part from two pieces of $\frac{3}{4}$" × $\frac{3}{4}$" × $\frac{1}{8}$" brass angle about 1" long, a piece of $\frac{1}{2}$" × $\frac{1}{8}$" brass strip, and a piece of $\frac{1}{4}$" square brass rod. Alternatively, you can machine the main body of the back cock from the solid and add on the square suspension support.

I will describe the fabrication method I use. Cut two pieces of brass angle of the size given above, each 1" long. File the ends square and mark out the curved shape of the feet of the cock, and also the position of the mounting holes which can be drilled now, $\frac{3}{16}$" (2 BA clearance). The shape of the feet is not critical — it is merely decorative and can be done however you please. Also mark out the vertical legs of the cock, $\frac{1}{2}$" wide, to meet the curves of the feet. Saw the waste away, leaving enough metal to clean up with a file later. Next cut a piece of $\frac{1}{8}$" brass plate, $\frac{1}{2}$" wide and 1" long. This now needs to be silver soldered to the top of the two angle pieces, with the joint indicated by the dotted lines on the drawing. I have made a simple jig to hold the pieces together because I make quite a few backcocks, but for one-off, the

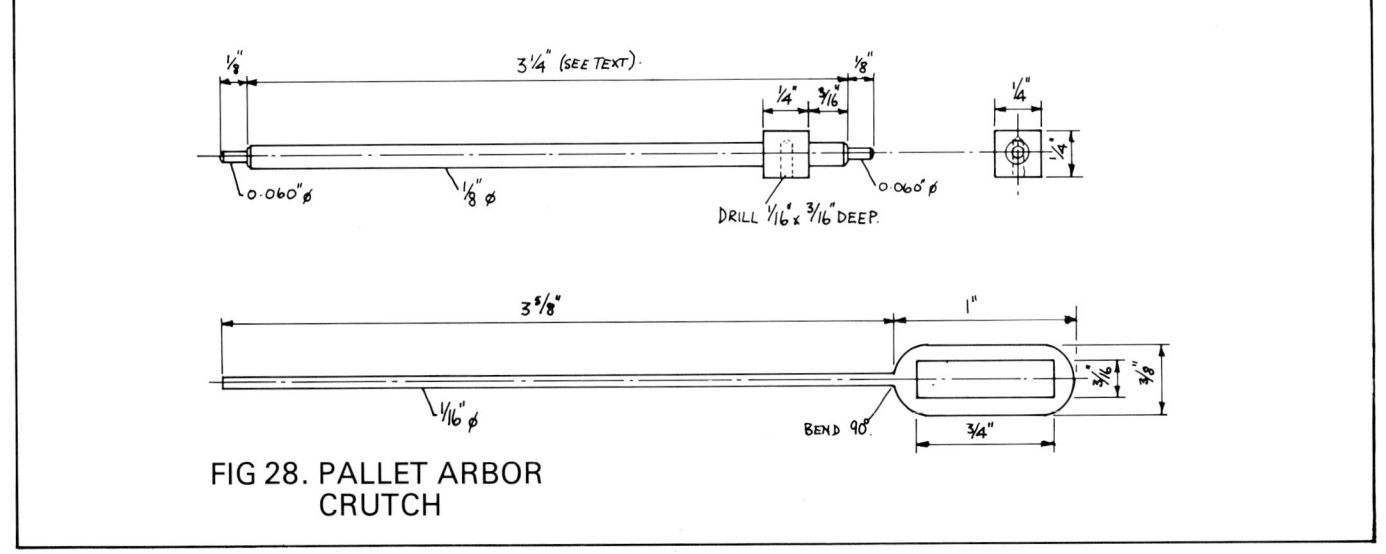

**FIG 28. PALLET ARBOR CRUTCH**

The pallet arbor and crutch are shown in Fig 28. The pallet arbor can be made in two ways. The first is to machine the whole arbor from a piece of $\frac{1}{4}$" square silver steel, leaving the boss into which the crutch is fitted — this is a straightforward turning operation either between centres, or in the four jaw chuck. By far the easiest method, however, is to fabricate this arbor from a length of $\frac{1}{8}$" diameter silver steel which can easily be turned to length and have the pivots finished, before soldering a small $\frac{1}{4}$" square block of mild steel with a central $\frac{1}{8}$" hole onto the arbor in the position shown on the drawing. The length of the arbor between the shoulders depends on your pillar length and the dimensions of your back cock, so it is as well to check this dimension from the job, not forgetting to allow the usual amount of endfloat.

The crutch is fabricated from a length of

37

FIG 29. BACK COCK

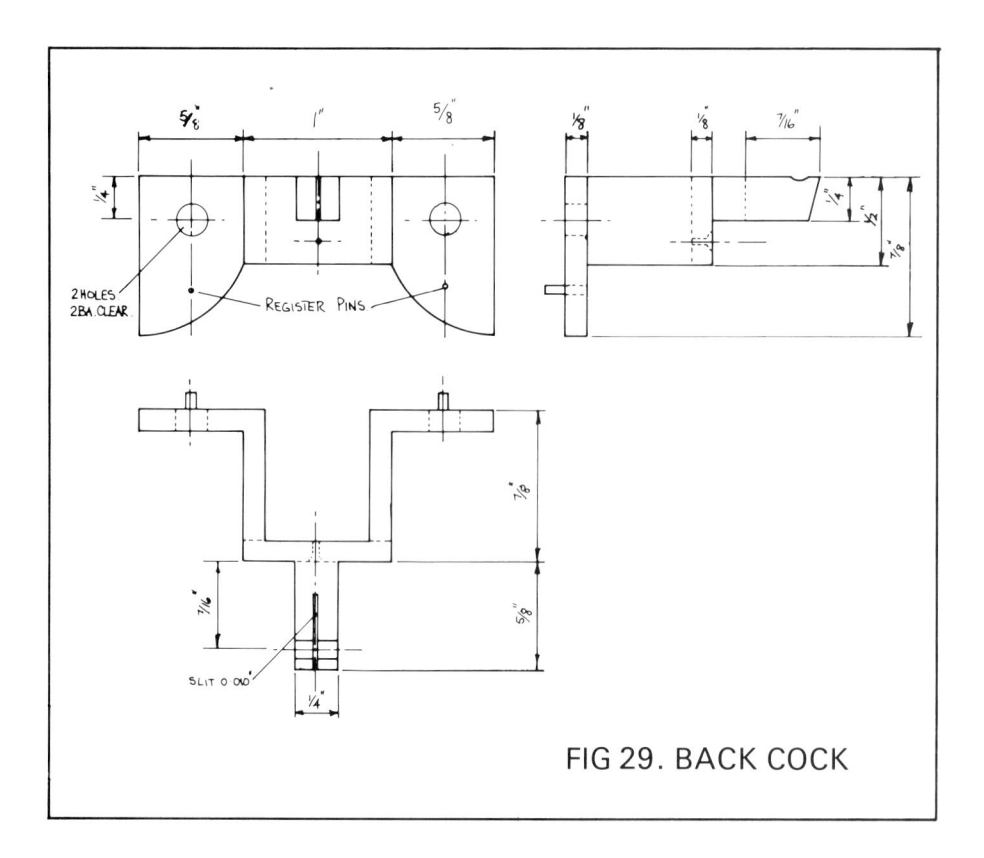

hole in the cock. Clamp the cock into position using a small pair of toolmakers clamps. Make sure that the top edge of the cock is parallel to the top edge of the plates. On my clock, the cock is about $1/16''$ down from the top of the backplate. With the cock clamped in position, check that the arbor can still revolve freely before spotting the backplate with a $3/16''$ drill through the mounting holes in the cock. Before removing the clamp, drill two holes in the positions indicated on the drawing. These should be drilled $1/16''$ diameter through the cock and the plate so that register pins can be fitted to the cock to ensure correct alignment. The pins are fitted in exactly the same way as you fitted the register pins in the plates at the beginning. With this done, take off the clamp, drill the plate 2 BA tapping size (No. 26), and tap both holes 2 BA. The cock is then fixed to the plate using two 2 BA cheesehead screws. The $1/8''$ hole in the backplate is now opened out to $1/4''$ diameter, and a slot cut from the top of the plate into the $1/4''$ hole. This slot should be wide enough for the pallet arbor to pass through freely when assembling the movement, and it can be seen clearly in the photograph.

*Plate 28. The escapement assembled between the plates.*

parts can be wired together easily enough. The $1/4''$ square piece has to support the weight of the pendulum so it is worth while making sure that this is really well attached. I usually fix it by turning a $1/8''$ diameter spigot on the end of the square rod, and I then rivet this spigot into a $1/8''$ diameter hole drilled in the cock, before silver soldering the whole lot together. After soldering, clean up with files, emery sticks etc. The pivot hole for the back pivot of the pallet arbor is drilled as indicated in the drawing, half way between the base of the square rod and the bottom of the cross piece. The hole is of course drilled undersize and broached out to fit the pivot.

The slit in the square rod is to take the pendulum suspension spring. I cut this on the milling machine using a $0.010''$ slitting saw, but it can just as easily be cut with a piercing saw. Take care to get the slit straight and square to the axis of the back cock, and also get it on exactly the same line as the pivot hole. The semi-circular groove across the end of the square support is simply filed using a round file. The spring fits through the slot and the brass pad locates in the groove so that the pendulum cannot slip and fall off the back cock. (Suspension springs are usually supplied with round brass pads riveted to the end.)

To locate the back cock on the backplate, it is best to fit the pallet arbor in the plates and locate the cock on the end of the arbor. To do this, after you have drilled the pallet arbor pivot holes through both plates with the plates pinned together as usual, open out the pallet hole in the back plate and ream it $1/8''$ diameter so that the pallet arbor is a good running fit in the hole. Assemble the plates and pillars, and then fit the pallet arbor through the hole in the backplate and into its front pivot hole. The arbor should revolve quite freely. Now place the back cock in position so that the rear pivot of the pallet arbor is located in its

The final operation on the pallet arbor is to fit the collet and pallets. The collet is riveted into the pallets, holding the collet on the split stake as with the wheels. I usually file a couple of small nicks in the hole in the pallets before riveting the collet, so that there is no danger of the pallets revolving on the collet. After riveting, run the reamer through the collet again to open the hole to size. It will help with setting up if the pallets are a stiff push fit onto the arbor, so don't run the reamer all the way through the collet.

Assemble the plates with the centre, third and 'scape wheels in position, fit the pallets onto their arbor and position them in roughly the right place to mesh with the 'scape wheel, and then fit the pallet arbor and back cock to the plates. You will now be able to position the pallets accurately on their arbor so that the 'scape wheel teeth engage in the middle of the pallet faces. When this is done, lightly mark the arbor as a reference for when you come to solder the collet to the arbor.

The pallets must be soldered on in the correct relationship to the crutch. This is the most difficult step to describe, and you must take care that it is done correctly! When the clock is running, the tick should be even as the pendulum swings each side of centre and releases a tooth on the 'scape wheel, i.e. the clock should be 'in beat.' On longcase clocks, beat adjustment is usually done by bending the crutch rod, but if the pallets are fixed to their arbor in the correct position and the movement is stood exactly level, there should be little, if any, bending necessary.

Basically, the pallets should release a tooth at each swing of the pendulum and the crutch should be offset an equal amount either side of its centre of motion when each tooth is released. Thus, if the bottom of the crutch is $1/8''$ to the right of centre when the entrance pallet releases a tooth, the same point on the crutch should be $1/8''$ to the left of centre when the exit pallet releases a tooth.

If your pallets are a nice tight fit on the arbor, you will be able to adjust their position relative to the crutch by simply revolving them on the arbor. If your collet is a loose fit and will not stay put, you will have to solder it in approximately the correct position and adjust by melting the solder each time you wish to alter their position.

To position the pallets correctly, stand the movement on a level surface with the back plate facing you and the crutch hanging vertically. At rest, the crutch should hang down the centre line of the plate. Next, from the back, revolve the centre wheel anti-clockwise using a finger to provide the power. Rock the crutch from side to side with power applied to the train and note the position of the crutch, relative to the centre arbor pivot hole in the back plate, when a tooth is released. If the pallets will not release a tooth, as they did in the depth tool, check the position of the back cock and the distance between the 'scape and pallet arbor holes, making sure that they are exactly the same distance apart as the runners were when the pallets were marked out in the depth tool. Also check that this distance is the same back and front. If all is well and the pallets do release the teeth, adjust the angular position of the pallets until teeth are released as described above, i.e. with the crutch an equal amount either side of centre. When you have positioned the pallets accurately, soft solder them to the arbor making sure nothing moves during soldering!

With the soldering done, replace the pallet arbor and back cock, and apply power to the centre wheel. This should cause the pallets to rock back and forth as they did in the depth tool. When setting up a recoil escapement, there are several points to look out for. First, examine the drops, or the amounts the 'scape wheel teeth move between being released by one pallet and arrested by the other. The amount of drop should be the same for both pallets, and there should be as little drop as possible consistent with free running of the escapement. Excessive drop is simply a waste of energy and also causes eventual pitting of the pallet faces.

If there is too much drop on the entrance pallet, (i.e. the pallet on the left looking from the front into which the teeth enter the escapement), then the pallets must be lowered slightly. This can be done by elongating the screw holes in the back cock, refitting the cock with the screws only finger tight and tapping the cock down slightly which will bend the register pins enough to make the adjustment.

If there is too much drop on the exit pallet (on the right looking from the the front, by which the teeth leave the escapement) the pallet faces need to be closed or squeezed in slightly. This can be done with care by squeezing in the vice. This is the reason why I only harden the pallet faces on an anchor, leaving the arms soft in case they need slight adjustment. Both these adjustments should be done in tiny increments (I nearly said minute, meaning small not 60 seconds!)

You will find that adjusting one pallet will affect the other and also affect the beat adjustment so mind how you go and remember what you do as you go along. If you find that you need drastic alterations, something is wrong with the bits you have made, so go back over what you have done and check the marking out of your pallets again.

Finally, check to see whether the 'scape wheel recoils after it has dropped onto the pallet faces. To do this, apply power to the train and rock the crutch from side to side with your fingers. When a tooth drops, continue moving the crutch in the same direction and watch the 'scape wheel carefully. If all is correct, the wheel should recoil or go backwards slightly after it has dropped onto the face of each pallet. If this is not the case, the pendulum will not receive any impulse to keep it swinging and the clock will slowly come to a stop as the pendulum swing dies away. If you have no recoil, again check back over your marking out to see where you have gone wrong.

I have discussed the setting up of the escapement at some length in order to solve any problems that may arise because of

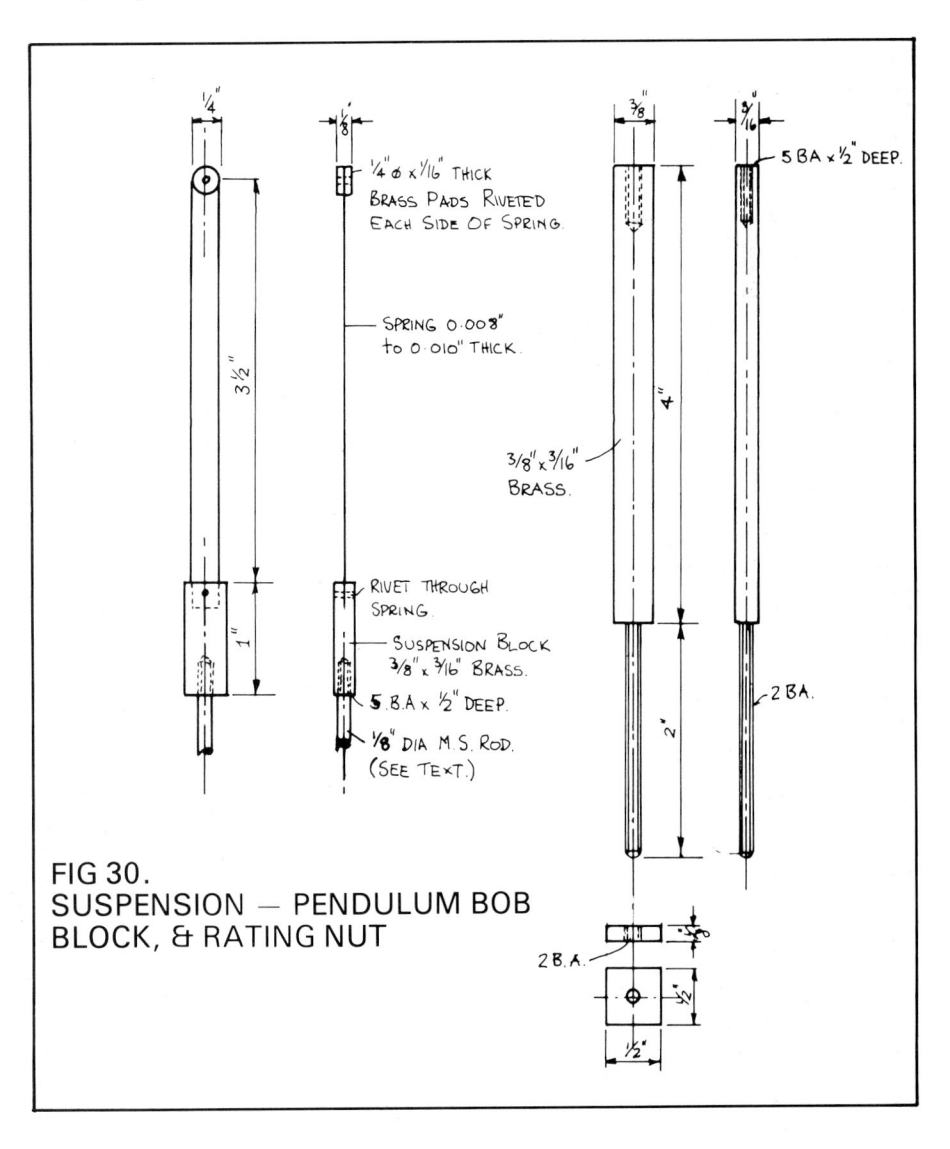

FIG 30.
SUSPENSION — PENDULUM BOB BLOCK, & RATING NUT

39

small errors in the making of the anchor. I must emphasise that these adjustments should not be necessary if the anchor was drawn properly and made accurately. Once I have positioned the pallets correctly on their arbor I usually expect a clock to go first time. Repairing a worn escapement can of course be 'a whole new ball game!'

**THE PENDULUM**

Figs 30 and 31 show the components which make up the top and bottom of the pendulum. These parts are connected together by a length of 1/8" diameter mild steel rod to give us the correct length.

Starting from the top of the pendulum we have the suspension spring and top block. To be honest, I usually buy these ready made from one of the usual suppliers of clock parts. Their price makes it hardly worth making suspensions, but I know some people wish to make every piece of the clock themselves so I have included details. Should you wish to buy the suspension, ask for a Longcase Suspension with two brass ends.

The spring is a length of spring steel, 1/4" wide and anything from 0.008" to 0.010" thick. A 4" length will be sufficient. The top end of the spring fits into the slot in the back cock, with the brass pads located in the groove filed across the top of the slot. The two pads, each 1/16" thick are parted off from a length of 1/4" diameter brass rod, after drilling 1/16" diameter in the lathe. These are then riveted to each side of the spring, using a steel clock pin as a rivet.

To do this operation, we need a hole in the spring which is hard. I was fortunate recently to acquire a pair of special pliers which will punch a variety of shapes and sizes of holes in clock springs. I am not sure if these are still available — they are certainly most useful and efficient. Before I obtained these pliers I made up a simple punch and die to make the holes. This is simply a steel or brass block with a 0.010" slit cut in it, and a 1/16" hole drilling through at right angles to the slit. The punch is a short length of 1/16" diameter silver steel rod with one end faced dead flat and hardened. To pierce the spring, simply insert it into the slit so that the position for the hole is in line with the hole in the block, fit the punch into the hole and give it a sharp tap with a hammer. This will produce a neat hole with no risk of splitting the spring.

The bottom suspension block is a 1" length of 3/8" × 3/8" brass which can be cut from the great wheels. File the material up all square. The only critical dimension is the 3/16" thickness. The block *MUST* be a free fit in the crutch slot WITHOUT SHAKE. It must on no account be tight, and any free play will result in loss of impulse to the pendulum. The 3/8" wide faces of the block should also be polished, as should the inner edges of the crutch slot. One end of the block is slit to take the spring which is again riveted like the top pads. The bottom end of the block is drilled centrally No.40 for 1/2" deep, and tapped 5 BA.

At the bottom end of the pendulum, there is a heavy bob, resting on a rating nut and sliding on a length of 3/8" × 3/8" brass. The rating nut is used for fine adjustment of the timekeeping of the clock. If the bob is moved up, the pendulum is shortened and the clock goes faster, and vice versa.

The pendulum bob block is self explanatory from the drawing. It is simply a length of brass bar tapped 5 BA × 1/2" deep at its top end, with a length of 2 BA studding silver soldered into its bottom end. The rating nut is also a simple part to make and dimensions are not critical.

The pendulum bob is another part that I often buy from the suppliers, complete with its block and rating nut. They are available either as lead filled brass shells or in cast iron with the slot for the block cast in. If you wish to make your own bob it is easier to make the brass type.

To do this, you must first make two brass shells from 20 SWG brass sheet. These can either be beaten on a sandbag or hollow wooden former and then planished, or you can spin them in the lathe. To spin them, first make a former from hardwood turned in the lathe to a cardboard template. The former can be held in the reverse or outside jaws of the 3 jaw chuck. Cut two discs of brass sheet a little oversize and anneal them by heating to dull red and allowing them to cool slowly. They are held against the former by a metal pad with one end faced off flat and the other centre drilled. Hold the disc centrally on the former, place the flat end of the pad against the disc, and tighten up the tailstock fitted with a revolving centre to hold the work firmly against the former.

The spinning tools are easy to make. The tool itself is a length of 1/2" diameter silver steel with one end shaped to a hemisphere and highly polished and the other end fitted with a handle for safety. This tool is used to force the revolving disc into contact with the former, so we need to fit a fulcrum into the toolpost to lever the tool against. This is made from a length of 1/2" steel square steel with a 3/8" diameter peg projecting from one of its faces near one end. This can be held in the tool post and moved to suit the job as it progresses.

To form the shape, revolve the blank as fast as possible, and lubricate it all over with thick grease or soap. Adjust the fulcrum so that you can start to force the blank into contact with the former from the centre outwards, and lock the saddle to the lathe bed. Run the tool out to the edge in gradual stages until the brass work hardens, and it will then need to be annealed again. Repeat the process until you have formed both shells into close contact with the formers. At this stage, their surfaces will be quite rough (unless you have done a fair bit of spinning and

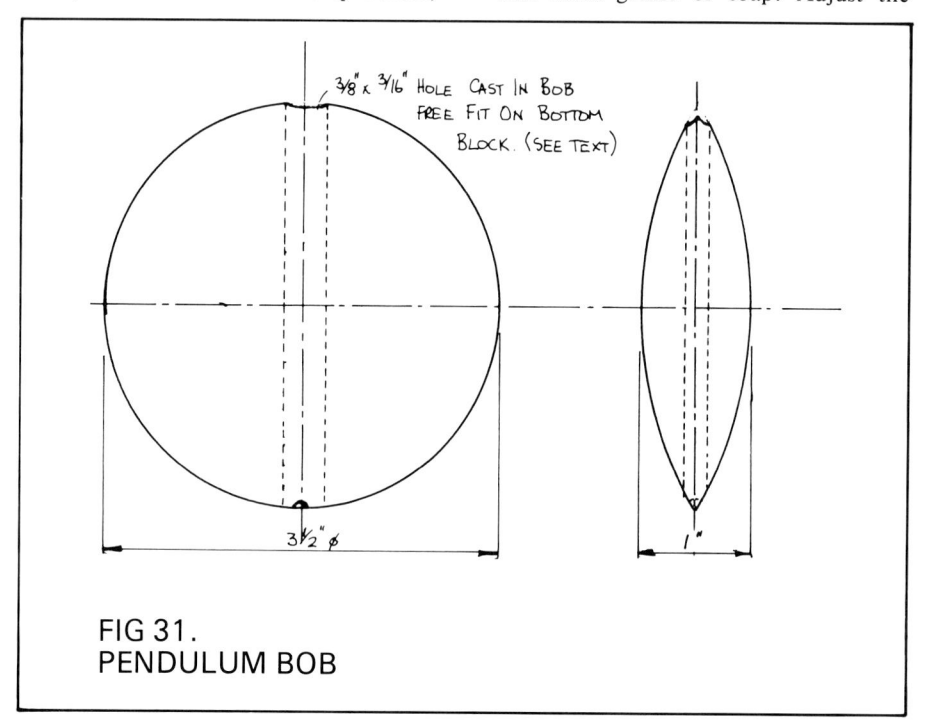

**FIG 31. PENDULUM BOB**

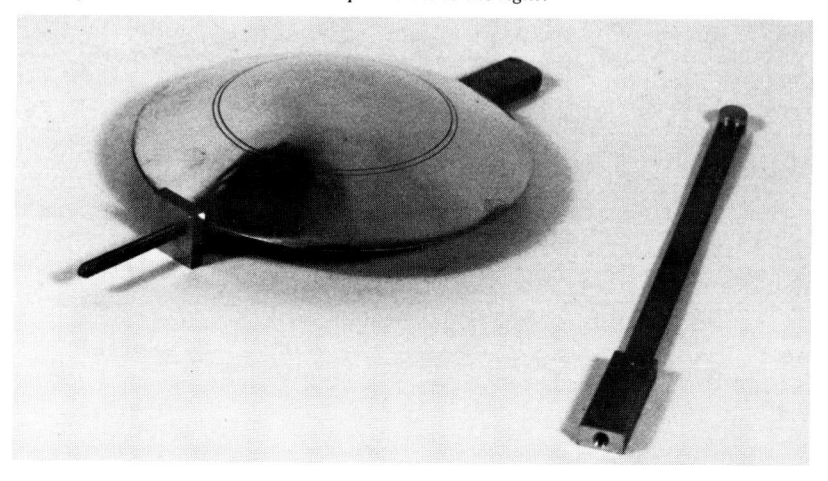

3/8" × 3/16" HOLE CAST IN BOB FREE FIT ON BOTTOM BLOCK. (SEE TEXT)

3 1/2" ⌀

1"

*Plate 29. The Pendulum for the Longcase Rack Strike Clock. The block and rating nut are in position in the bob and the suspension is to the right.*

know what you are doing!) They can be cleaned up later with emery cloth followed by a good polishing on the buff.

With the shells done, mark their centres and with dividers scribe a 3½" diameter circle to mark their outer edge. Cut to these lines with tinsnips and file the edges smooth. Drill the centre of one shell about ½" diameter — this will be the back shell of the bob so choose the one with the worst surface. The hole is needed later to pour the lead into the shell. The next stage is to silver solder the two shells together. Prepare the joint faces by rubbing the shells on a sheet of coarse emery cloth to produce a flat joint face on each half. Keep trying the two halves together until a perfect joint is obtained. Flux both faces and stand one shell convex side down on a piece of tube or a tin can. Place the other shell on top so that the edges line up, and place a weight on it to hold it in position. Warm it all up and run 'easy-flo' silver solder round the joint leaving no gaps. Let it all cool down before pickling in acid.

We now need to fill the bob with lead, leaving a rectangular hole across the diameter to take the brass block. To do this we need to make a core piece to fit into the shell and it is easiest to use aluminium for this as the lead will not stick to it. A piece of dural or aluminium alloy ⅜" × ³⁄₁₆" and about 4" long will do nicely. On one end of this, turn a ³⁄₁₆" diameter spigot, holding the metal in the 4 jaw chuck, to represent the 2 BA rating screw. File a small nick across the bottom of the shell joint line until you break through the wall of the shell, and then open out the hole with a round needle file until the spigot on the core piece will fit in tightly. Mark a diametrical line across the shell and at the opposite side from the hole already made, cut out a slot in the rim of the bob to take the rectangular part of the core piece. You can now fit the core through the rectangular hole and locate the spigot in its hole at the bottom of the bob, so that it passes through the centre of the bob.

The bob is now ready for filling with lead, which can be poured in through the hole in the back shell. The usual method is to make a funnel from thin card (a postcard is ideal), which is placed in the hole. Melt the lead in an old saucepan or something similar and pour it down the funnel in a steady stream until it almost fills the funnel. This forms a reservoir to allow for shrinkage on cooling and also leaves a spigot to hold the bob in the lathe to polish the front face. I should warn you to make quite certain that the inside of the bob is absolutely dry before pouring in the lead. Molten lead and moisture definitely do not like each other, the consequences can be very dangerous, so make sure and pre-heat the bob shell to avoid danger.

When the lead has cooled, it should be possible to knock out the core, leaving a neat hole through the bob. Hold it by the lead spigot in the lathe and remove all marks from the front face using progressively finer emery cloth. Polish the face on the buff to a high scratch free finish and finally saw off the lead spigot.

You can now try your brass block through the bob. It will probably need drawfiling on all faces to make it fit smoothly. The aim is for an easy sliding fit with no shake, so that the bob will slide up and down its block during adjustment.

As I mentioned earlier, I usually buy in pendulum bobs. They are not very expensive, but as you can imagine from the above description, they take quite a lot of making! The diameter of the bob is not important. They are available in 3½" and 4½" diameters. A lot of people think that it is the weight of a pendulum which controls the rate of a clock. This is not the case. The length of the pendulum is the important thing.

The final part of the pendulum is simply a length of ⅛" diameter mild steel rod. Cut a length 34" long and thread both ends 5 BA for ½" long. Screw on the suspension tightly at one end, and the bob block at the other, so that the wide faces of the two blocks are at right angles to each other. Slide the bob onto its block and run the rating nut on about half way up its thread. This will do for a start — the timekeeping accuracy can be adjusted later.

**THE WEIGHT PULLEYS**

These are shown in Fig 32 and we need two, one for each train. The pulley is turned from either a slice of 1¾" diameter brass bar or a disc cut from a piece of ³⁄₁₆" thick Compo brass sheet. Drill and ream the centre hole ¼" diameter and then turn the diameter holding the blank on a ⅜" diameter mandrel with a ¼" spigot. Turn one face down and form the recess, then reverse on the mandrel to do the other face. The dimensions of the recess are not critical. The groove round the rim which takes the line is formed using the same tool bit that was used to groove the barrels. On very good quality clocks, especially precision regulators in glass cases, the makers often went to considerable trouble to cross out the pulleys to make a decorative feature of them. You may do this if you wish, but as the pulleys are not usually seen when the clock is cased, I don't think there is much point in doing so.

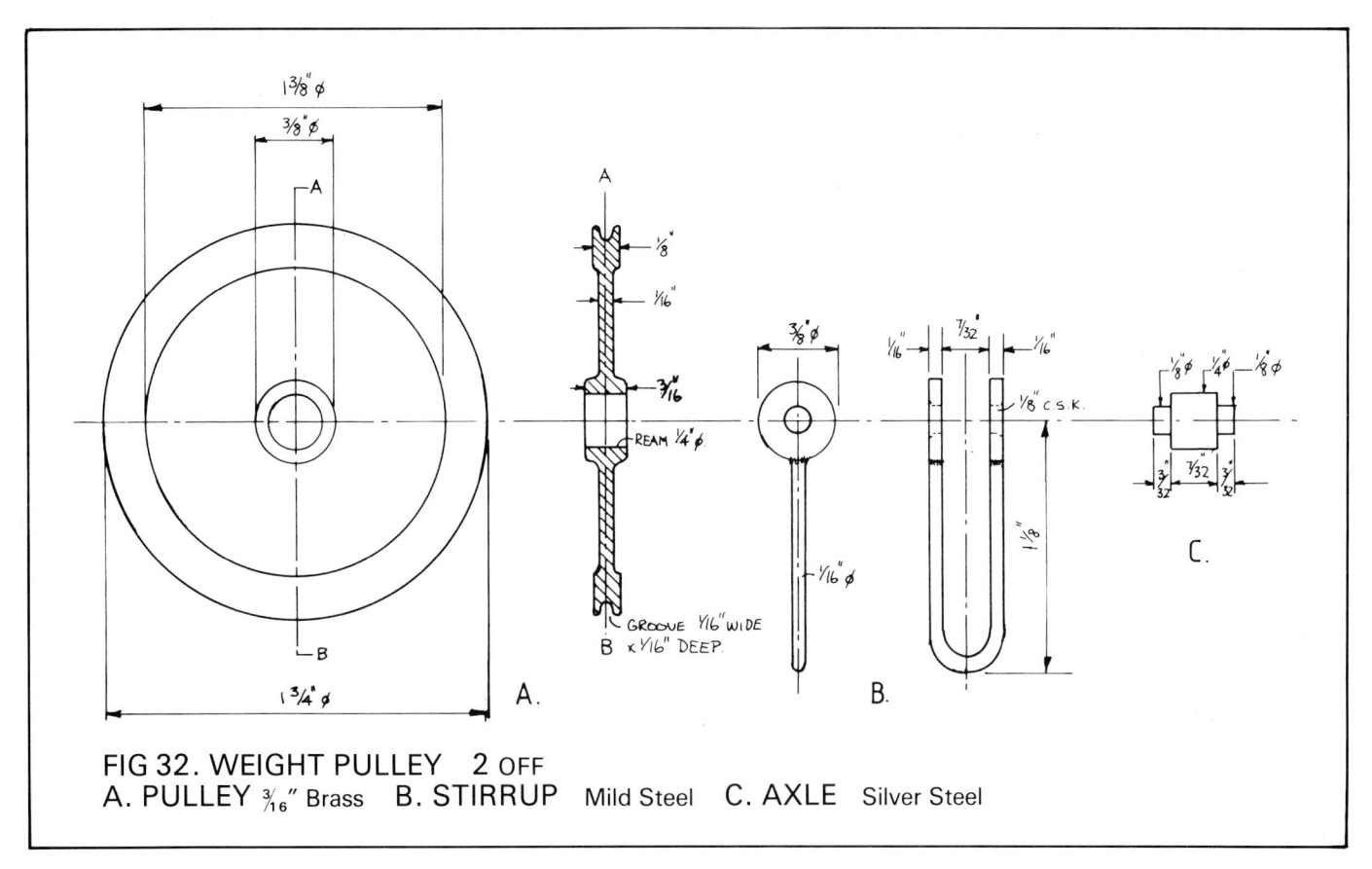

The stirrups can be made in two ways. The first is to draw a flattened out development of them and then cut them in one piece from ¹⁄₁₆" thick steel sheet. The U-shaped part can either be filed to make it

FIG 32. WEIGHT PULLEY   2 OFF
A. PULLEY ³⁄₁₆" Brass   B. STIRRUP Mild Steel   C. AXLE Silver Steel

round, or you can leave it square. Again, it does not make any difference. If you make one piece stirrups, take care to get the holes in line after bending.

The easiest way to make these parts, however, is to drill a length of ⅜" diameter mild steel rod in the lathe using a ⅛" drill and then part off four slices each ¹⁄₁₆" thick. Then bend up the U-shapes from ¹⁄₁₆" diameter steel rod and silver solder the discs to them. If the discs are held on a piece of ⅛" rod with a spacer between them, the silver soldering can be done quite easily and the holes will be accurately lined up. After soldering, clean up all over with emery and polish before final assembly.

The axle is another simple turning job. The only points to watch are that the ⅛" spigots are concentric, that the ¼" diameter lengths are a nice running fit in the pulleys, and that the shoulder lengths of the ¼" portions are slightly longer than the thickness of the pulleys. With all the parts made, fit the axles into their pulleys, spring the stirrups over the ends of the axles and rivet the lot together with light hammer blows checking that the pulleys revolve freely.

You will now be able to set up the time train of your movement and get it running. For now, you can hang anything on the pulley for a weight. The usual antique longcase clock has lead or cast iron weights of about 12 pounds each. However, I think that this is excessive and only accelerates wear, especially on the barrel arbor holes. My clock and another made by a friend will run on about 7 pounds on the time side and slightly more on the strike train. It is best, therefore, if you experiment a little to find how much weight you need to make the clock run reliably. To find this, you can hang a tin on the pulley and fill it up with offcuts or bits of lead.

### THE WEIGHTS

I always buy weights ready made — plain lead ones being the usual on longcase clocks with wooden trunk doors. If you don't live near a supplier, the postage is rather expensive and so you may wish to make your own. You have three options open to you. Firstly, and probably the easiest, is to use pieces of mild steel bar about 2½" diameter. This size bar weighs about 1.4 lb per inch, so 6½" will weigh about 9 lb. You should by now know how much weight your clock needs, so you will be able to calculate the size of your weights easily enough.

Standard lead weights are available weighing 7, 10 and 12 lb. A twelve pound lead weight is about 2½" diameter by 6½" long, so again you can work out what size you will need if you wish to cast your own. Tin cans make convenient moulds and these can easily be cut off when the lead has solidified. If you cast your own weights, bend up hooks from ⅛" diameter steel rod and cast them in situ.

The final method for making weights is to make brass cases from thin wall tube with sheet brass tops silver soldered to the tubes. You can ignore the weight of the cases, or put each case on a pair of scales and add lead to the scales until you have the weight you need. Fit hooks to the tops of the cases, support them rigidly with the open bottoms facing upwards, melt your lead and pour it in. You will not need to fit bottoms to the cases as these are not seen, and the lead will key itself to the inside of the case.

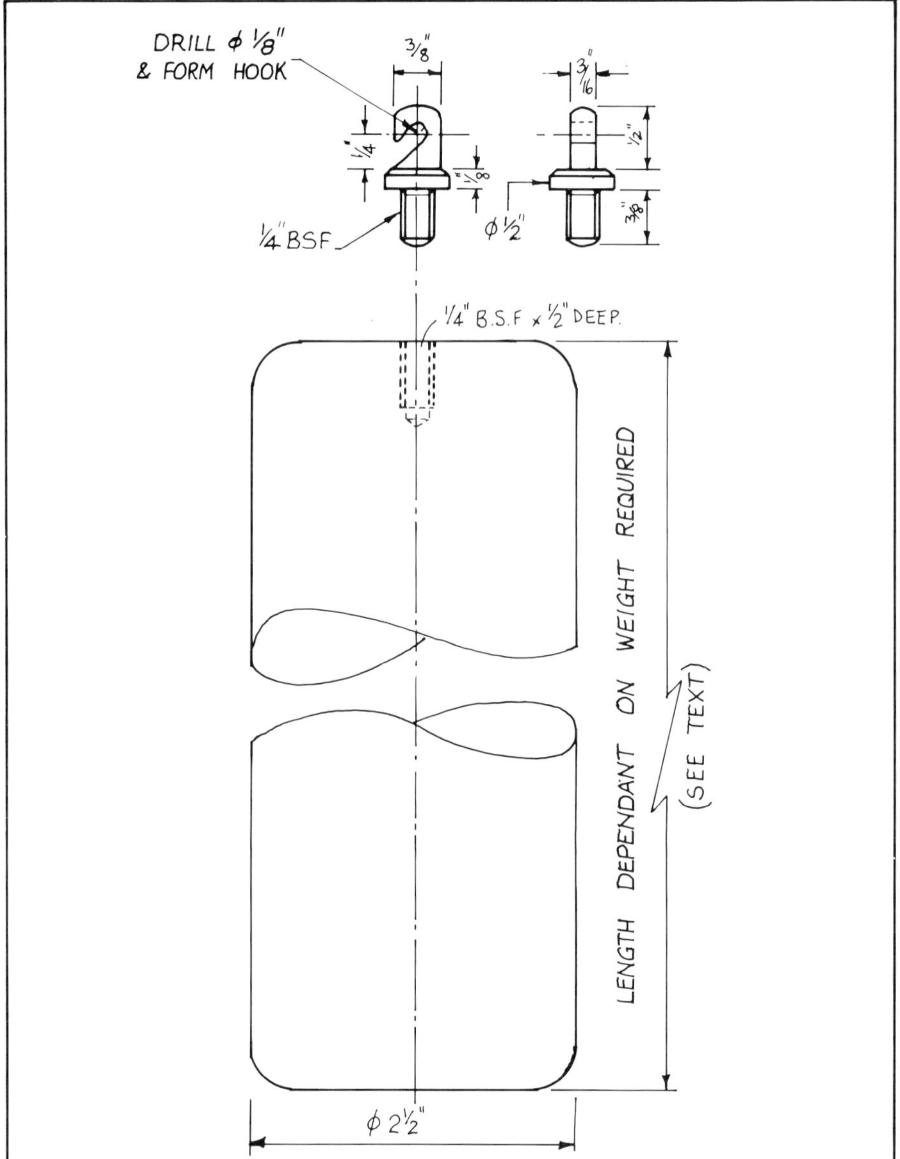

### SETTING UP TO TEST THE TIME TRAIN

I include this short section now so that you can run the time train whilst the rest of the motion and strike work are made. It is nice to get a movement ticking and this keeps up the enthusiasm!

You will need some sort of test stand or 'horse' about 4 feet high on which to stand the movement. This will need a top about 12" × 8" with holes cut in for the lines to pass through, and a slot cut for the pendulum to pass through. I have two or three horses which are free standing, but you can easily fix up a small shelf on a spare bit of workshop wall (if you are lucky enough to have any spare wall space!). Fig 34 shows the layout of a suitable seatboard. Points A and B are two ³⁄₃₂" diameter holes to pass the lines back up through the seatboard after they have gone round the pulleys. Their position is not important. The most important point, however, is that the seatboard is set up level, both side to side, and fore—aft.

I use Perlon lines for the weights, as this is far less liable to damage through kinking etc. These lines are available from suppliers listed in the appendix. Ask for a set of Perlon Longcase Lines. They come in one length and have to be cut in half. They are plenty long enough and you will probably have to shorten them when you finally fit the movement into its case. Leave them full length for now.

To fit the line into the going barrel, grip the end ¼" or so in a pair of pliers and bend it. Feed the line through the small hole at the end of the groove and twist the line to encourage its bent end to pop out of the hole in the flange. Pull a couple of inches through and knot the end tightly. Cut off the excess, leaving about ¼" spare, and heat the end in a flame to form a blob to stop the knot slipping. Then pull the knot into the barrel, making sure the end does not stick out and interfere with the Great Wheel.

**FIG 33. WEIGHTS AND HOOKS**
WEIGHTS 2 OFF  ⌀ 2½" Mild Steel
HOOKS 2 OFF  ⌀ ½" Brass

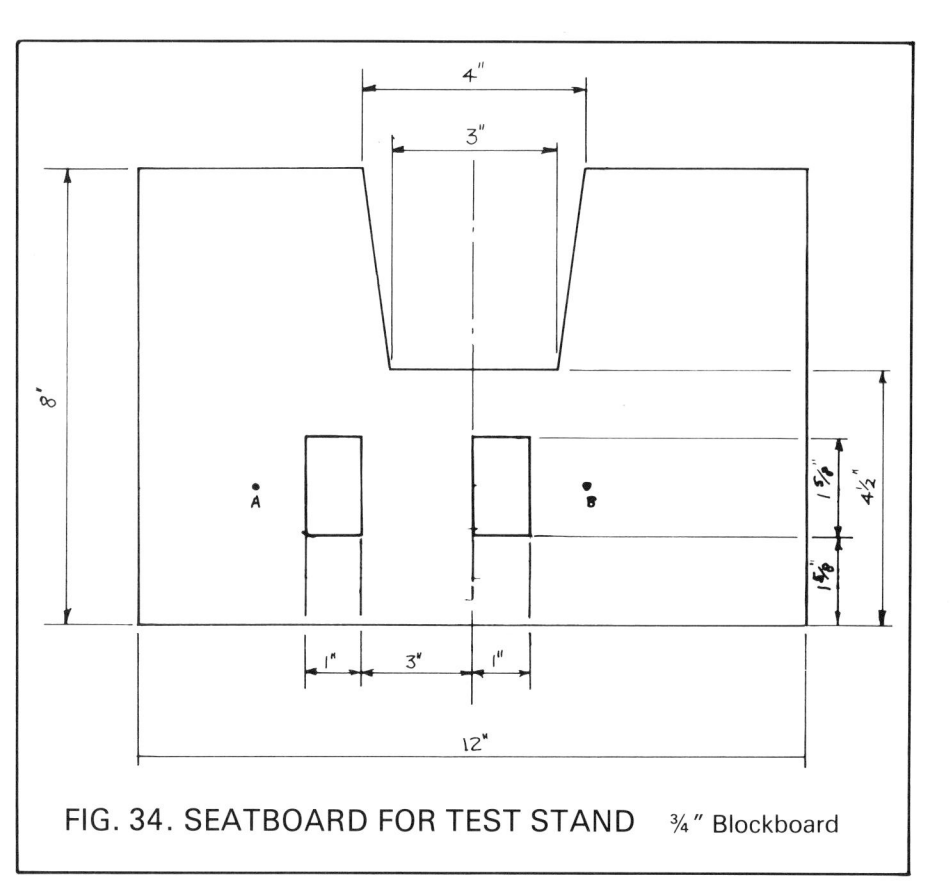

**FIG. 34. SEATBOARD FOR TEST STAND**  ¾" Blockboard

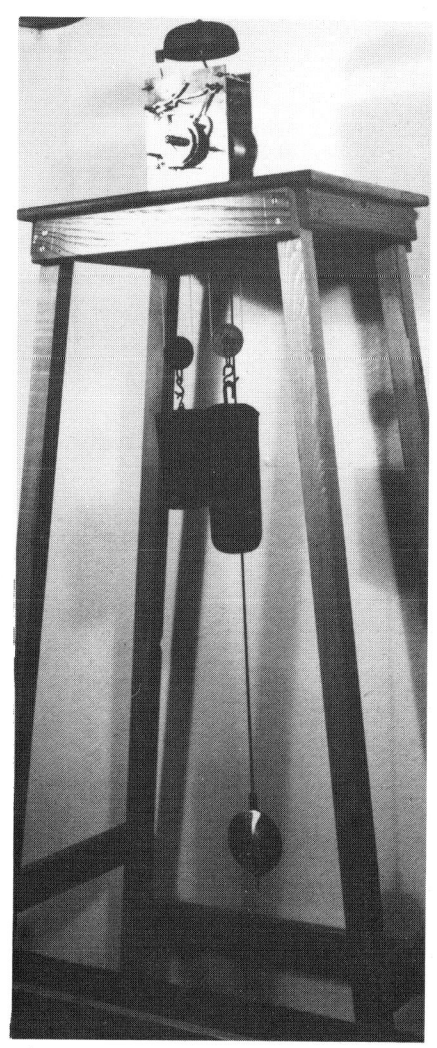

*Plate 30. Finished movement set up for testing on horse.*

Make sure all the pivot holes are clean and assemble the going train in the plates. Before fitting the pallets, apply a tiny drop of thin oil (proper clock oil is best), and make sure the train revolves freely. Put a drop of oil on each pallet face and fit the pallet arbor, again applying a drop of oil to its pivots.

Feed the line down through the rectangular hole in the seatboard, thread on a pulley and pass the line back up through the small hole in the board. Wind the going barrel up, guiding the line into its groove as you go along. You can do this by winding the barrel with your fingers for now — I will describe the key later. When the barrel is full, pull the spare end of the line up until the pulley is an inch or so below the seatboard. You can now wrap the free end of the line round a bit of rod, and tie it off so that the rod supports the line over the small hole.

Hang your test weight on the pulley and, if all is well, the crutch should start to oscillate madly! Don't let it do this for long as it could cause damage to the 'scape wheel teeth.

Hang the pendulum on the back cock, passing it up through the crutch on the way. Apply a drop of oil to both sides of the crutch slot where the bottom suspension block passes through it. If you now swing the pendulum and you have set up the escapement as described, the clock should at last start ticking!

Listen to the tick-tock to determine if it is even. If it isn't, stop the pendulum and let it hang free. When it has settled down, position an indicator point at the bottom end of the rating screw and then move the pendulum bob to one side, or the other, until a tooth is released. Note how far you have moved it. Then move the pendulum in the opposite direction until the next tooth is released, and again note the distance the end of the screw is from centre. You will find that the swing needs to be shorter on one side. If you now grently bend the bottom of the crutch rod towards the short side of the swing, you will be able to get the clock in beat. Only a tiny amount of bending is necessary, and you will probably overdo it and have to bend it back again. You will soon see what happens when you try it for yourself. When you have got it right, sit down and watch it for a while. One second ticks are very soothing!

# Chapter 8

**THE MOTION WORK**

Most clocks have their hour and minute hands mounted concentrically on the dial. The exception to this is the precision regulator which, to avoid friction, has separate hour, minute and second dials engraved on a common dialplate, with the hands mounted directly on the ends of their respective arbors. The mechanism which allows us to have concentric hands revolving in the same direction is called motion work, and this is fitted on the fromnt plate of the movement behind the dial. I described the motion work in some detail in Chapter 1, and I would advise readers to look back and read this again, and refer to Fig 2 to refresh their memory.

The two minute wheels have 39 teeth. The reason for this rather odd number is that we have two pairs of gears, with each pair mounted on common arbors to provide a reduction of 12:1. It is necessary, therefore, to find two sets of gears which will give an overall ratio of 12:1, are the same Module for convenience, and which will mesh correctly at the same centre distance. This is usually achieved by making the ratio between the two minute wheels 1:1, and the ratio between the reverse minute pinion and the hour wheel 12:1, using 39/39 and 72/6, as:

$$\frac{39+39}{2} = \frac{72+6}{2}$$

If you can not obtain 39 divisions with your dividing equipment, it is quite alright to cut 40 teeth on the minute wheels, but use blanks calculated for 39 teeth. All this will do is give you slightly thinner teeth.

Fig 33 shows the two minute wheels, the friction spring and the reverse minute arbor. Commence with the minute wheel and pipe. I have shown the two minute wheels as $\frac{3}{32}''$ thick. The only reason for this is that as you push the minute pipe back against the spring when you pin the minute hand into place, the minute wheel moves back relative to the reverse minute wheel. If these two wheels are made slightly thicker you will have a little more latitude with your fitting. If you have $\frac{1}{16}''$ blanks, it will be necessary to just take a little extra care to ensure that on final assembly these two wheels line up properly. Also you will have to adjust the thickness of the square on the reverse minute arbor after a trial assembly.

As we do not want too much backlash in the motion work, the wheels should be meshed together a little tighter than usual, and so take particular care when cutting these wheels to ensure that they are exactly concentric otherwise you will have problems later. You will no doubt be pleased to learn that none of the motion wheels are crossed out.

The minute pipe is a straightforward piece of turning. The centre hole is drilled and reamed either $\frac{3}{32}''$ or $\frac{1}{8}''$, depending on which type of centre arbor you decided to make. The pipe should be a nice running fit on the arbor. At one end of the pipe, turn a $\frac{3}{16}''$ diameter spigot to be a good tight fit in the minute wheel. The wheel is soft soldered to the pipe. It is important to remember that if you are using the simple depthing tool, leave all wheels off their pipes, etc, until all the depthing is complete.

In passing, I must mention that I have received several letters asking about the use of 'Loctite' in such applications. I do use this very useful adhesive at times and so far, I have found it to be very good, provided that the instructions are followed carefully — particularly with respect to cleanliness. I have used 'Loctite Superfast Retaining Compound' to fix pinions, barrel flanges, fusées, etc, so far without trouble. I cannot obviously vouch for its

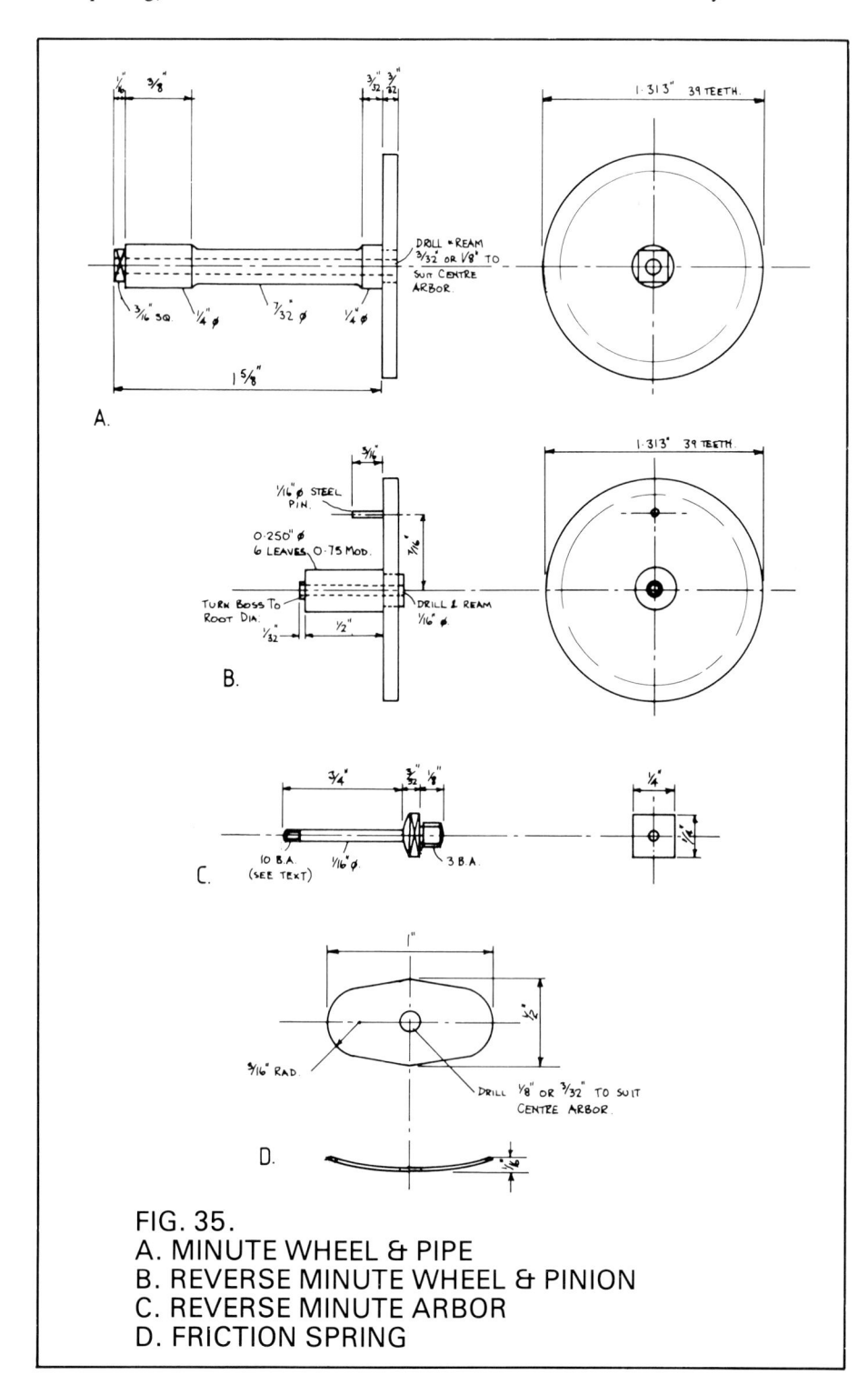

**FIG. 35.**
A. MINUTE WHEEL & PIPE
B. REVERSE MINUTE WHEEL & PINION
C. REVERSE MINUTE ARBOR
D. FRICTION SPRING

long term strength.

The easiest way to form the square on the other end of the pipe is to mill it with the work held in the dividing head. This can also be done on the lathe, using a milling spindle on the vertical slide. The 7/32" diameter length of the pipe is not at all critical.

The reverse minute wheel and pinion assembly is again a simple component. The pinion is made of brass and has six leaves, 0.75 Module. If you make this pinion 0.250" diameter, it will mesh nicely and will leave enough room down its centre for a 1/16" reamed hole. Again, the wheel is soft soldered to a 3/16" diameter spigot turned on the end of the pinion. Allow about 1/64" extra on the length of this spigot so that a small boss is left behind the reverse minute wheel as shown on the drawing. The pin fitted into this wheel is simply a tapered clock pin tapped into a No. 54 hole in the wheel, and then trimmed to length. This pin will eventually trigger the striking mechanism.

The friction spring is made from a piece of 20 or 22 SWG brass sheet. After cutting it out and drilling the hole, bend the spring to the shape shown and then hammer it gently to harden it.

The reverse minute arbor screws into the frontplate after depthing the two minute wheels. The arbor is made from 1/4" square mild steel rod, to the dimensions shown. As we will have to make a few of these studs, it is worth making a simple fixture to hold square rod accurately in the three jaw chuck to save fiddling about centring the four jaw chuck! A fixture can be simply made from a 1" length of 7/16" diameter mild steel rod with a hole drilled down its centre — the hole should be the same size as the distance across the corners of the square rod. After drilling in the lathe, slit the tube along its length. You can now push a length of 1/4" square rod into the tube, then grip the tube in the three jaw chuck. This will squeeze the tube and grip the square rod concentrically in the chuck.

I make these arbors in two parts. The square rod is turned down and threaded 3 BA; then using a thin parting tool, undercut the root of the thread at its shoulder so that it will screw fully home in the frontplate. Then drill a 1/16" hole down the centre. The arbor end is parted off to length and then held by the thread in the three jaw chuck. (If you wish you can make a simple mandrel out of a piece of rod with a 3 BA threaded hole, to hold this part in the chuck with no risk of damaging the thread). Using either a form tool or a fine file, slightly dome the face of the square as in the drawing. The 1/16" arbor is simply a length of silver steel soldered into the square. The 10 BA thread is to take a nut to hold the reverse minute assembly onto its arbor. Care should be taken not to thread too far down the arbor — the aim is that with the nut screwed on tight, the wheel and pinion should revolve freely with a tiny amount of endshake. Traditionally, the assembly was held in place by a tapered pin through the end of the arbor, but I can tell you that it is no fun trying to drill a cross hole through a 1/16" diameter rod, so do it my way!

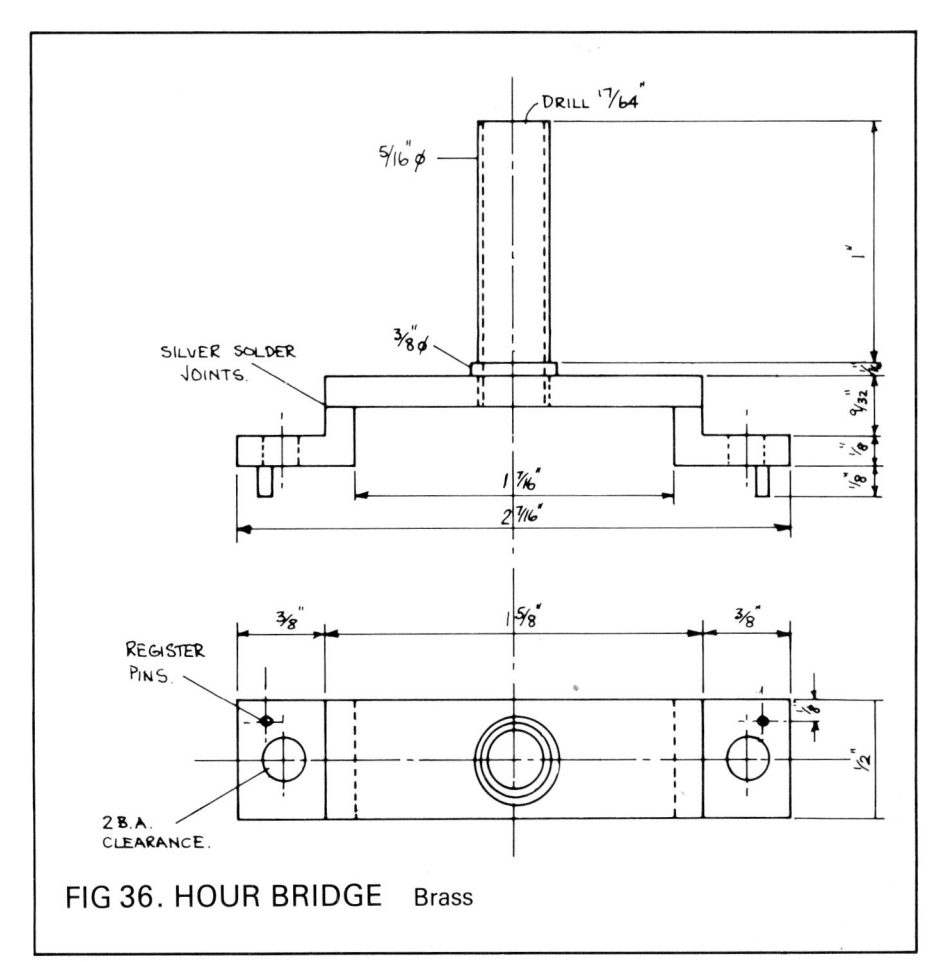

## THE HOUR BRIDGE

Like the back cock, the hour bridge is fabricated from brass angle and strip. The purpose of the bridge is to provide a bearing for the hour wheel and pipe, concentric to the minute pipe. The latter passes through the tube mounted on the hour bridge which is drilled to clear the minute pipe. The hour pipe is a good running fit on the outer diameter of the tube on the bridge.

The 'legs' of the bridge are made from two pieces of 1/2" × 1/2" × 1/8" brass angle. These are sawn to length and their edges filed up to leave them nice and square and 1/2" long. One leg of each piece then needs to be shortened so that the height of one face of each piece of the angle is 9/32", measured from the outer corner of the angle. This can be done by sawing and filing, but I prefer to mill them both together to ensure that both legs are identical in height.

Cut a piece of 1/2" × 1/8" brass strip slightly over 1 5/8" long to form the top deck of the bridge. Once again, a very simple jig will make the silver soldering operation much easier. This jig is simply a length of steel or brass strip, about 3/4" × 1/4" and 3" long. Mark a centre line along the length and on this line, 2 3/32" apart, drill and tap two holes, 2 BA. If you now drill the 2 BA clearance holes in the legs of the bridge, you will be able to position the legs accurately the correct distance apart and exactly parallel to each other, screwed down to the jig. Clean and flux the joint faces as usual, before positioning the top piece of the bridge on the legs. A few turns of soft iron binding wire will hold everything in place for the soldering. I usually stand the whole lot upside down on a flat firebrick, heat it all up and apply the solder to the inside corners under the bridge. If you are very careful and don't get too much silver solder on the joints, there should be very little cleaning up to do. Before you fit the

**FIG 36. HOUR BRIDGE** Brass

*Plate 31. Minute Wheel, Reverse Minute Wheel and Arbor.*

tube, file the base of the bridge up all over, keeping all the surfaces flat and true, and all the corners nice and sharp. Finish to a smooth scratch free state by rubbing all surfaces on a sheet of fine emery cloth or wet and dry paper held taut on a sheet of glass, or using emery sticks.

The tube is another simple turning operation. The hole can be left as drilled, because it is not a bearing surface and the minute pipe should not touch the inside of the tube on assembly. Hold a length of ⅜" diameter brass rod in a collet and turn the ⁵⁄₁₆" diameter portion 1" long. This should be brought to a very fine finish to be a good running fit in a ⁵⁄₁₆" reamed hole. Centre drill and drill about 1⅜" deep, gradually opening the hole out to the finished ¹⁷⁄₆₄" diameter. Part off a 1¼" length and hold the ⁵⁄₁₆" portion in a collet to turn the other end down to ⁵⁄₁₆", leaving a ¹⁄₁₆" long collar at ⅜" diameter. The ⁵⁄₁₆" spigot at this end should be a good push fit in a ⁵⁄₁₆" reamed hole.

The final stage with the bridge is to mark out the position of the central hole to take the tube. Drill and ream the hole ⁵⁄₁₆" diameter, making sure that the hole is exactly square to the bridge in all planes. The tube must stand absolutely vertical in all planes when the bridge is screwed to the front plate. The tube is finally fitted either by soft soldering or using 'Loctite'. The steady pins are fitted later on assembly.

### THE HOUR PIPE

This component is shown in Fig 37. The hour hand fits on the ⅜" diameter spigot, whilst the hour wheel fits on the ½" diameter spigot at the opposite end, held in place with a slip washer fitted into the ¹⁄₁₆" wide groove. The two 8 BA holes in the collar of the pipe are to mount the snail later, and these are drilled from the snail when it has been finished.

The 1⅛" diameter collar is another optional part of the clock. If you intend to fit datework, which I will describe as an extra later, then this collar needs to have 30 teeth of 0.8 Module to drive the datework — so you may as well prepare for this now — you can cut the teeth later if you wish. If you don't intend to fit datework, the collar can be 1" diameter.

I save a lot of turning and waste material by making the pipe from a length of ½" diameter brass rod, and the collar from a piece of ³⁄₁₆" compo brass sheet. This method also makes it easier to cut the teeth.

A disc of brass about 1¼" diameter

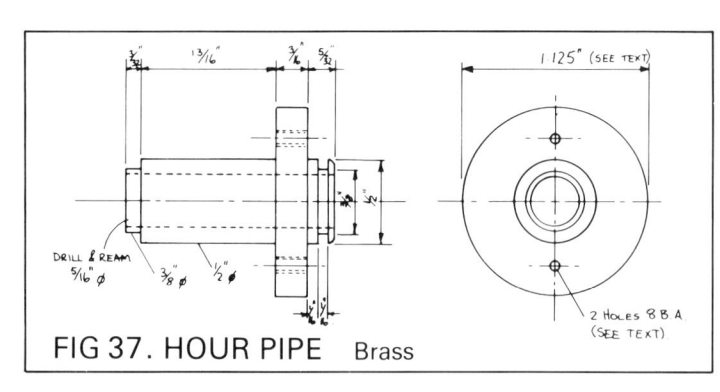

**FIG 37. HOUR PIPE** Brass

is drilled and reamed through its centre ½" diameter. Hold the disc on a mandrel, as for wheelcutting, turn to 1.125" diameter, and cut the teeth as already described. The pipe is simply a length of ½" diameter brass, drilled and reamed ⁵⁄₁₆", with a ⅜" diameter spigot turned on one end as shown. Leave the groove for the slip washer until later. The collar is then soldered or loctited to the pipe in

Plate 32. Hour Bridge.

the correct position.

The hour wheel and slip washer should be made next from ¹⁄₁₆" compo brass. The hour wheel has 72 teeth, 0.8 Module, and a ½" diameter centre hole to fit the pipe. Do not drill the wheel for the 10 BA hole yet. The slip washer is also easily made from brass sheet. Turn it to diameter on a mandrel in the lathe, then drill the centre hole out to ⅜" diameter and drill a ½" hole ⁷⁄₁₆" off the centre. Saw and file to form the keyhole shape as shown. Again, leave the hole for the screw until later.

The hour wheel can now be fitted to its pipe and the position of the slip washer groove marked. Hold the long end of the pipe in a ½" collet and turn the groove using a fine parting tool. Adjust the width and depth of the groove until the slip washer fits tightly. We want to be able to revolve the hour wheel on its pipe so that we can adjust the position of the hour hand — thus making setting up easier — but the hour wheel must not be able to move on the pipe of its own accord. When you are satisfied with the result, drill through the slip washer No. 55 (10 BA tapping size) in the position shown, assemble the wheel and slip washer on the pipe, and with the No. 55 drill dimple the hour wheel to mark the position of the hole. Take it all to pieces again, drill the wheel right through and tap 10 BA. Open out the hole in the slip washer to just clear the head of a 10 BA cheesehead screw. Screw a 10 BA screw into the hour wheel and tighten it, then cut off the waste thread at the back of the wheel and rub it flush with the wheel on a sheet of fine emery cloth. Remove the screw and reassemble the wheel on the hour pipe, fit the slip washer and also the screw which will prevent the washer sliding off.

### FITTING THE MOTION WORK TO THE FRONTPLATE

Fig 39 shows the positions of the holes for the hour bridge and the reverse minute wheel arbor, relative to the vertical centre line down the front plate. The reverse minute arbor is the first component to be fitted. For this, mark a line very lightly on the front plate, 1" to the right of the centre line of the plate and above the centre arbor position. Depth the two minute

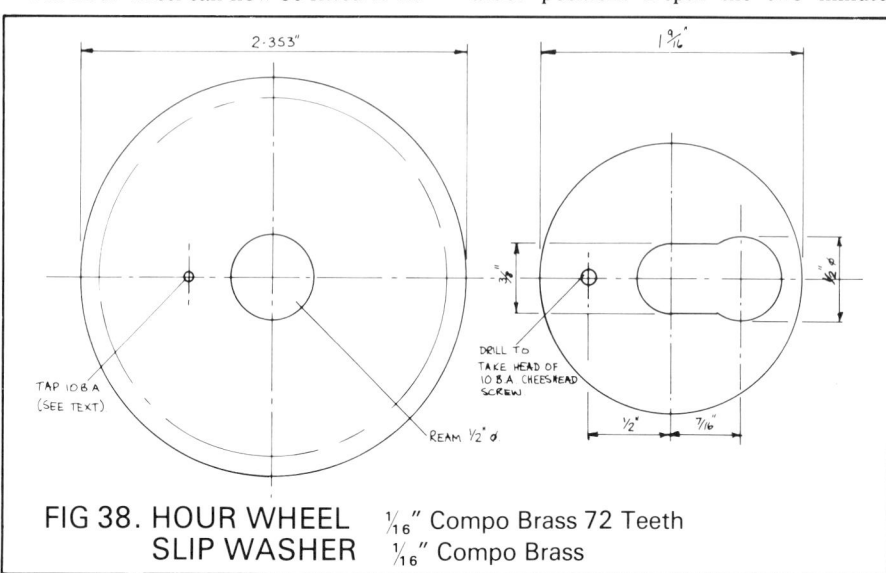

**FIG 38. HOUR WHEEL** ¹⁄₁₆" Compo Brass 72 Teeth
**SLIP WASHER** ¹⁄₁₆" Compo Brass

*Plate 33. Hour Pipe, Hour Wheel and Ship Washer.*

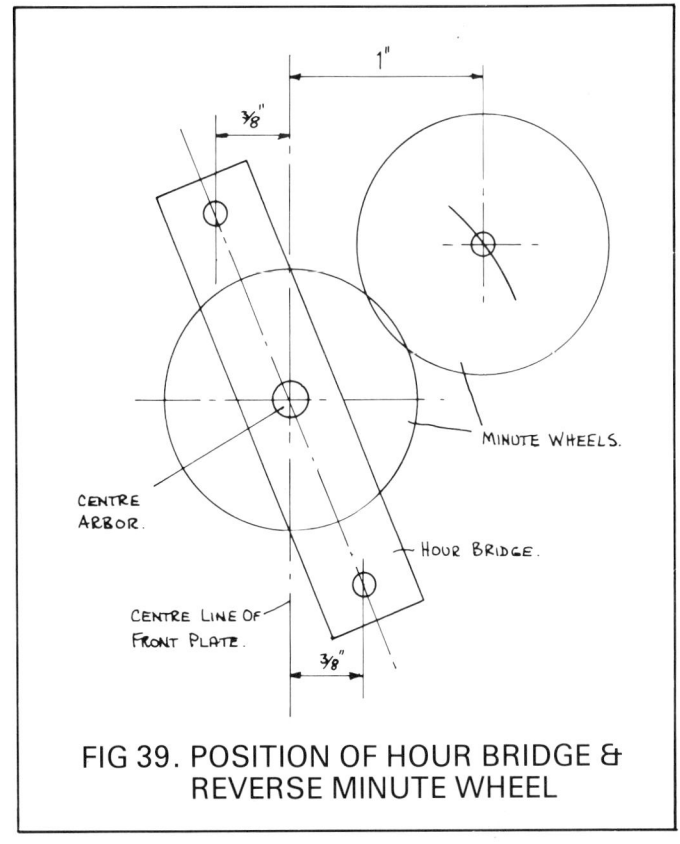

**FIG 39. POSITION OF HOUR BRIDGE & REVERSE MINUTE WHEEL**

wheels in the simple depth tool. To do this, you will need to turn up a little bush which is a tight fit in one of the wheels with a $3/32''$ centre hole to fit the runner of the depth tool. Mount the wheels on the tool and adjust the runners for correct mesh. We want the two wheels to mesh as closely as possible to eliminate backlash — but not so close that there are stiff spots as the wheels revolve. Take your time with this operation and get it correct. Some backlash is inevitable, but try to get as little as possible. When you have it right, scribe an arc on the plate, with the cone pointed runner positioned in the centre arbor hole, and the arc cutting the line just scribed 1" to the right of centre.

Without adjusting the depth tool, mount the hour wheel and the reverse minute pinion. In order to do this, you will need to make a ½" bush with a $3/16''$ centre hole to mount the hour wheel, and you will also have to make a new runner to fit the depth tool, but with the blunt end turned down to $1/16''$ diameter to mount the pinion. The wheel and pinion should mesh nicely without adjusting the centre distance. If they don't mesh correctly, you will have to adjust the depth of both pairs of gears to obtain the best compromise, but if you have made all the parts as described they should fit well. When you are satisfied that all is well, check the marking out on the plate before very carefully centre punching the point where the arc crosses the line to the right of the centre. Drill this point and tap the hole 3 BA to take the arbor which can now be screwed into position.

Mount the minute wheels onto the pipe and pinion, and fit the centre arbor between the plates. Slide the friction spring over the centre arbor, convex side towards the plate, and then fit the minute wheel and pipe over the centre arbor so that the back of the wheel bears on the upturned ends of the spring. The reverse minute wheel is then fitted on its arbor so that the two wheels mesh together. Check at this stage that they spin freely.

In order to position the bridge, mark two lines ⅜" either side of the centre line — one to the left of centre above the centre arbor hole, the other to the right of centre below it. With the minute pipe in position on the centre arbor, fit the tube of the bridge over the minute pipe, and centre the bridge to the minute pipe, while at the same time positioning the bridge at an angle on the plate so that the lines ⅜" either side of centre appear centrally through the 2 BA clearance holes in the bridge feet. Clamp the bridge in this position with toolmaker's clamps and, after re-checking that the bridge is central to the minute pipe, spot through the mounting holes in the bridge to mark the hole positions on the front plate. Take everything apart and drill the two holes No. 26 and tap them 2 BA.

Re-assemble the components again and fit the hour pipe assembly to the bridge. You should now be able to revolve the minute wheel which should drive the reverse minute wheel and thus the hour wheel without any tight spots, and with little backlash.

*Plate 34. Front of Movement showing Motion Work in place.*

47

# Chapter 9

## THE RACK AND SNAIL

As I have described earlier, the number of strikes at each hour is controlled by the Rack and Snail. If you refer back to Chapter 1, and read the appropriate text again, you will refresh your memory on how these items operate.

The snail is made from a $\frac{1}{16}''$ thick brass blank, 2.353" diameter. The centre hole is drilled and reamed $\frac{1}{2}''$ diameter. The outer edge of the snail is divided into twelve steps, each step occupying 30 degrees of the disc with an exactly equal drop from one step to the next. From the drawing, you will see that the drop from the largest diameter step to the smallest is $\frac{19}{32}''$ or 0.594". If we divide this by 11, we find that each step is 0.054". The procedure for marking out the blank is as follows.

**FIG 40. SNAIL**
$\frac{1}{16}''$ Brass  Bore $\frac{1}{2}''$  O.D. 2.353"

*Plate 35. Marking out the divisions on the snail blank.*

Coat one side of the blank with layout blue, and mount it on a $\frac{1}{2}''$ mandrel on your dividing attachment so that you can get twelve divisions. Using a scribing block set to centre height, scribe a line across the diameter of the blank then revolve the blank $\frac{1}{12}$ of a turn, and repeat the process until all 12 divisions are marked. Now transfer the mandrel complete with blank to the lathe. Mount a scriber in the toolpost so that its point is at lathe centre height, facing the blank, and adjust the cross slide so that the tip of the scriber is exactly lined up with the edge of the disc. Zero the cross slide collar, or make a note of the reading on the dial. It is now an easy job to advance the cross slide 0.054", bring the scriber into contact with the blank, and scribe an arc across one segment of the snail revolving the blank by hand. Repeat this process until all the steps are marked out.

I usually make snails two or three at a time, and so I find it quickest and most accurate to mill the steps on a rotary table.

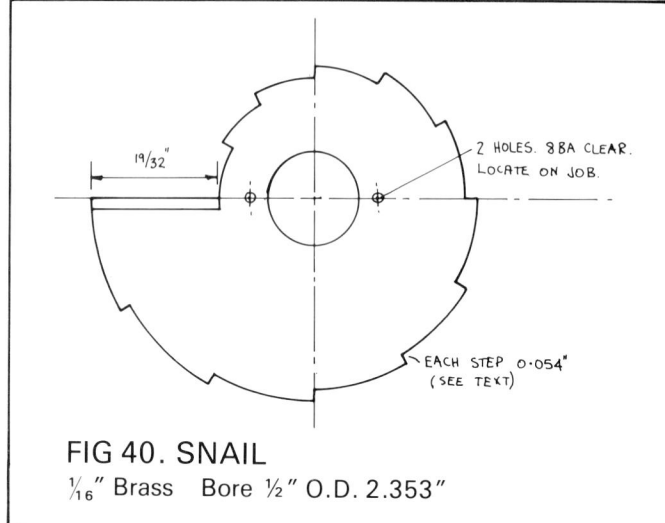

*Plate 36. Marking out the snail steps.*

All that needs to be done with this method is to feed the cutter in 54 thou, rotate the blank $\frac{1}{12}$ of a turn and so on. This produces very accurate snails easily. However, for one off, it is probably not worth setting up to do this. In this case, after you have marked out the blank accurately, very carefully saw round the steps with a fine piercing saw almost on the line, leaving a tiny amount to clean up with fine files and emery cloth. The vital point with the snail is that all of the steps down are exactly equal. The actual size of the step is not absolutely critical as long as they are all the same, as adjustments can be made to the rack tail later. However, if you keep closely to the dimensions given, then you should have no trouble setting up later.

The snail is fixed to the hour pipe with two small screws as shown in the drawing. Traditionally, these parts were riveted together, but this makes future cleaning very difficult and so I always use screws wherever possible.

The rack is the next part to make, and this is shown in Fig 39. The difficult part of this job is cutting the teeth, since the part is difficult to hold. The only way this can be done on a machine is to make a fixture to hold the blank. The blank is a piece of 1/16" thick mild steel sheet. I usually start with a triangle as indicated on the drawing by dotted lines. The rack is shown printed actual size so that you can trace the outline which is very difficult to dimension. Centre punch the blank in the position indicated and scribe an arc 2 7/8" radius across its top end. This will be the radius of the tips of the teeth. Saw and file away most of the waste, leaving a little to machine accurately to the finished size. Drill and ream the centre point of the arc 1/4" diameter.

You will now require the fixture illustrated in the photograph, Plate 38. I made mine from a piece of 1" diameter mild steel bar and a piece of steel plate 3/8" thick that I had in stock. Anything similar will do. Turn a 1/2" diameter spigot 3/4" long on the end of the bar, and thread it either 1/2" BSF or BSW. Undercut the thread at its root with a small parting tool so that you can screw the plate hard up to the shoulder to ensure that it is square to the axis of the bar. Finally, turn a 1/4" diameter spigot on the end of the threaded portion, leaving a length of thread the same length as the thickness of the plate you are using. Mark the position of the hole on the plate and scribe an arc 2 3/4" radius. Saw off the waste and file almost down to the line. Drill and tap the hole to suit the threaded spigot and screw the plate hard onto the bar. I use a little 'Loctite' to ensure a permanent fixing. Now, hold the bar in the three jaw chuck and set the lathe into a slow back gear speed to turn the arc on the end of the plate to the line previously marked. This radius will be the root diameter of the rack teeth.

Fit the rack blank onto the spigot on the fixture, and drill two holes in the waste part of the blank and through the plate so that the blank can be fixed to the plate with two small nuts and bolts — 2 BA is a good size for this job. Fix the blank in

*Plate 37. The completed snail ready to be fixed to the hour pipe.*

position, and return the fixture to the lathe where you can now turn the arc on the blank to exactly 2 7/8" radius. You can measure this dimension as you go along using a vernier caliper across the arc and the spigot on the fixture. As the spigot is 1/4" diameter, simply deduct 1/8" (the radius of the spigot) from your vernier reading — so the finished measurement should be indicated as 3", when the arc is 2 7/8" radius.

The next stage is to cut the teeth on the

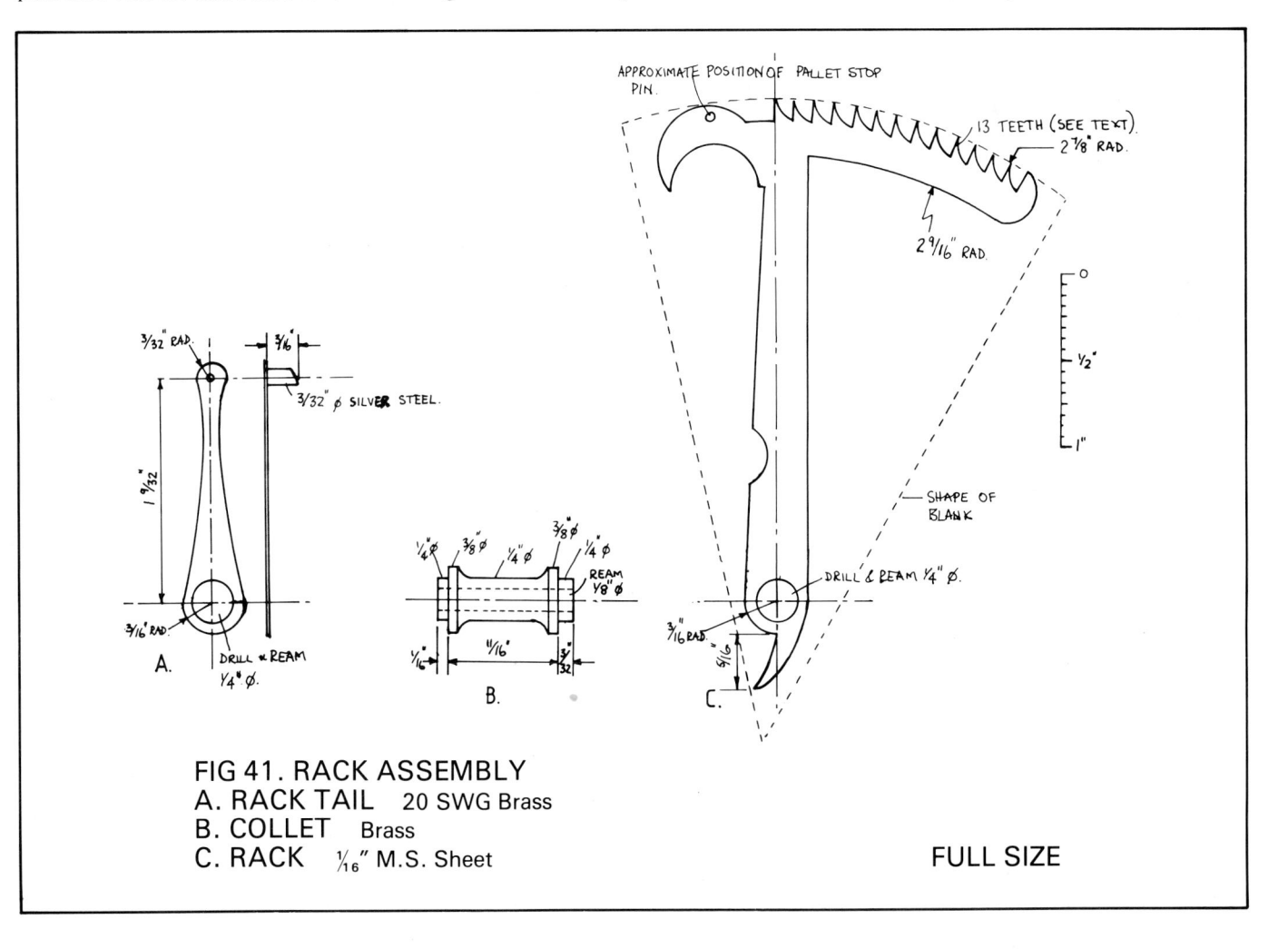

FIG 41. RACK ASSEMBLY
A. RACK TAIL   20 SWG Brass
B. COLLET   Brass
C. RACK   1/16" M.S. Sheet

FULL SIZE

*Plate 38. The set-up on the milling machine for cutting the rack teeth.*

blank. To determine the position of the teeth on the blank, trace the outline of the rack from the drawing, and mark out the shape on the blank. This will give you the position of the radial flanks of the first and last teeth for setting up purposes. We need to cut 14 spaces which will give us 13 teeth, and we need to make a special flycutter from a piece of ¼" diameter high speed steel.

The distance between the tips of the teeth is for all practical purposes ⅛", and the tooth spaces are ⅛" deep. As you will see from the drawing, the teeth are a similar shape to the 'scape wheel teeth — with one radial flank and one curved flank — and so the cutter is almost the same as the one made for the 'scape wheel. As an easier alternative, you can cut teeth with one radial flank with the other at an angle of 45°, using a standard milling cutter if you already have one of these. Some antique clocks have curved rack teeth, some have V shaped teeth as just described — so the choice is yours. If you use a 45° cutter, the teeth will automatically be ⅛" pitch when they are ⅛" deep.

To cut the teeth, the blank must be held on its fixture in your dividing attachment, in either the lathe or the milling machine, set up as for wheelcutting. In order to make the most convenient set up with your equipment it is as well to set up for cutting before you make the cutter, so that you can decide which way round to grind the curved edge of the cutter. The dividing attachment needs to be set up to cut 144 teeth to give the correct pitch. Centralise the point of the cutter as for wheelcutting, and then line up the blank using your marking out of the first or last tooth on the rack. The cutter speed needs to be fairly slow on steel, and it is best to experiment a little with your set up to find the speed which cuts easiest and leaves the best finish. Use plenty of cutting oil, and feed the cutter in very slowly. The procedure is exactly the same as wheelcutting. Feed the cutter down in stages, cutting two adjacent teeth alternately until the tip of the first tooth is formed. Again, do not make the tooth tip a sharp point but leave a slight flat for strength. If you have made your cutter properly, its tip should just touch the arc machined on the holding fixture when the tooth tip is correctly formed. When you have got the set up right, go ahead and cut all the teeth — 14 cuts in all, making 13 full teeth.

When the teeth have been cut, saw the outline of the rack, and clean up to the lines by filing. The small semi-circular depression on the left hand side of the leg of the rack is to clear the strike winding arbor when the rack drops fully over into the twelve position. Do not file this out yet — leave it until you have made a trial assembly of your movement to see exactly where and how much clearance you need. The gathering pallet stop pin should also be left for now. The position of this depends on the finished dimensions of the pallet.

The rack is mounted on a collet which revolves on a pin screwed into the front plate. The collet is a simple turning job and should present no problems. The only point I shall make here is that you should make the ¼" diameter spigots at each end of the collet a good tight push fit into the rack and the rack tail. These parts will eventually be soft soldered together, but you will need to adjust the assembly when you fit it to the movement so that you get the rack and tail at the correct angle to each other. It is a considerable advantage if you are able to adjust this angle, and then remove the assembly from the clock for soldering, without the parts moving in the meantime. If they are loose, I can guarantee a most frustrating time trying to get them right!

Now we come to the rack tail. If you have made the snail and all the other parts of the rack exactly to the dimensions given, then the rack will work properly if that too is made to the sizes on the drawing. From experience, however, I know only too well that a slight variation in one or two dimensions can throw the whole of the geometry out, and so I will explain the theory in order to help anyone who has a problem.

There are several methods of working out the geometry involved. The method I show is the most convenient and the most straightforward. The following method is described in *Britten's Horological Hints and Helps*, (a most useful book which I recommend to anyone interested in clocks). The construction is shown in Fig 42.

Point A is the centre of motion of the rack. From this point, draw two lines, AB and AC. AB passes through the first tooth of the rack, when it is at rest, and AC passes through the tooth which the rack

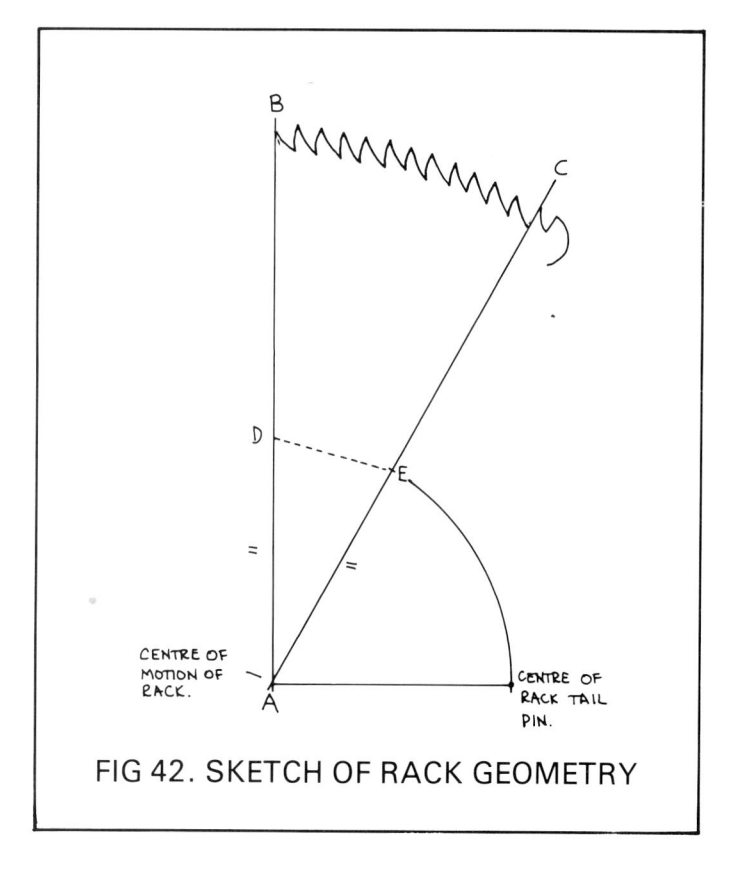

FIG 42. SKETCH OF RACK GEOMETRY

**FIG 43. A. RACK SPRING**
**B. RACK PIVOT    3 OFF Mild Steel**

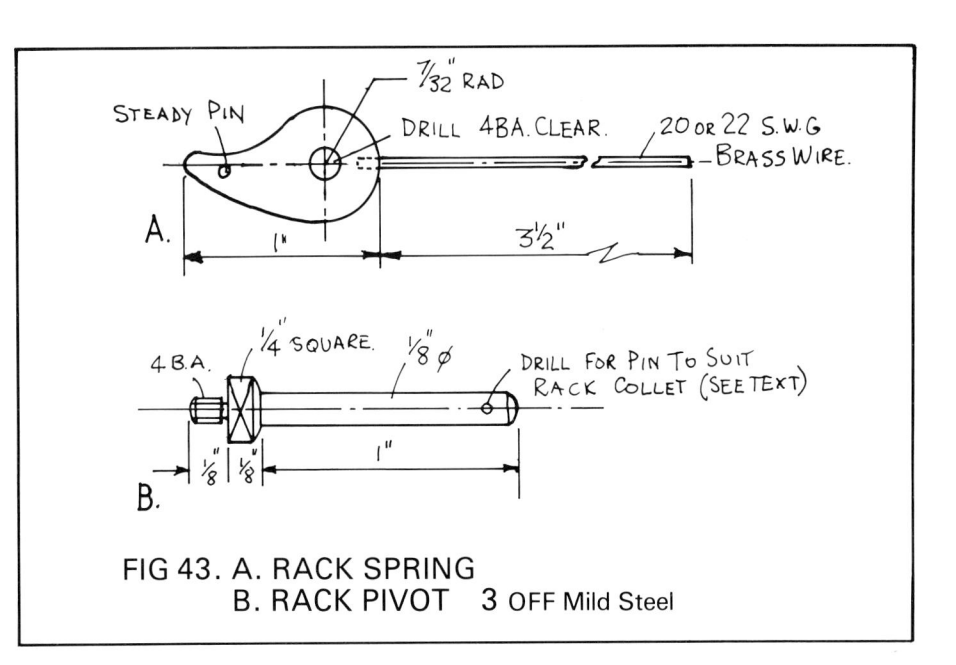

must fall to strike twelve. Thus, the angle between the two lines is equal to the angle through which the rack moves when it is released to strike twelve. Now measure the distance between the maximum diameter of the snail (the one o'clock step) and the minimum diameter (the twelve o'clock step). In our case, this should be $19/32"$. We will call this dimension DE. Set a pair of compasses to $19/32"$ (DE) and with the point of the compasses on line AB, find the points on the lines AB and AC that are exactly the distance DE apart so that AD and AE are equal. The distance AD or AE gives us the distance between the centre of motion of the rack tail to the centre of the pin fitted into the tail which stops the rack falling when it meets a step on the snail. On my movement, this dimension is $1\,9/32"$.

The rack tail is made from a piece of 20 swg brass sheet to the dimensions given. Remember to make the hole in the tail a good stiff fit onto the collet so that you can set up easily later. The pin which engages with the snail is made from $3/32"$ diameter silver steel, and is riveted to the tail. You will note that the end of the pin is bevelled off, as is the 'leading edge' of the snail. This is done so that in the event of the rack failing to drop at sometime, the pin on the rack tail will ride up over the snail, and so prevent the clock being damaged or being stopped. To make this possible, file the waist of the tail so that it is thin enough to bend upwards, but not so thin that it becomes flexible in the sideways plane as this would make the striking unreliable. Do not solder the rack assembly yet. We need to make a couple of parts to set it up correctly.

## THE RACK SPRING

When the rack is released at the warning, it drops with the assistance of a spring bearing on the hooked part at the bottom of the rack. The spring is shown in Fig 43, and its position on the movement in Fig 2. The mounting foot of the spring is made from a piece of $1/16"$ thick brass sheet, filed to shape, and drilled for a screw and steady pin as shown. The spring is a $3\frac{1}{2}"$ length of thin brass wire, about 20 or 22 swg. The wire is soft soldered into a hole drilled in the foot, leaving the wire straight. It is far easier to determine the correct shape of the spring when it is in place. After soldering the spring to the foot, rest the wire on a steel block, and tap it over it's full length with a hide mallet to work harden it and make it springy.

## THE RACK PIVOT

This is shown in Fig 43. It is turned up from $1/4"$ square mild steel in exactly the same way as the reverse minute wheel arbor — but to different dimensions. The cross hole for the retaining pin should be positioned so that when the rack collet is placed on the pivot, and the retaining pin is inserted, the collet is free to revolve, and has about $1/64"$ end play. In total, we shall require three pivot pins, all $1/8"$ diameter. Of the three, the rack pin is the longest, so make three identical while you have the set-up, and shorten the other two as required later. You can of course either turn the pivots from solid, or fabricate them as I have described earlier.

As the location of the rack pivot is dependent on several other parts which have yet to be made, I will leave this until the next chapter when we will make the rest of the levers and the gathering pallet, plus the bell hammer and spring.

*Plate 39. The completed rack assembly. Also shown are the arbor and the spring (both before and after bending).*

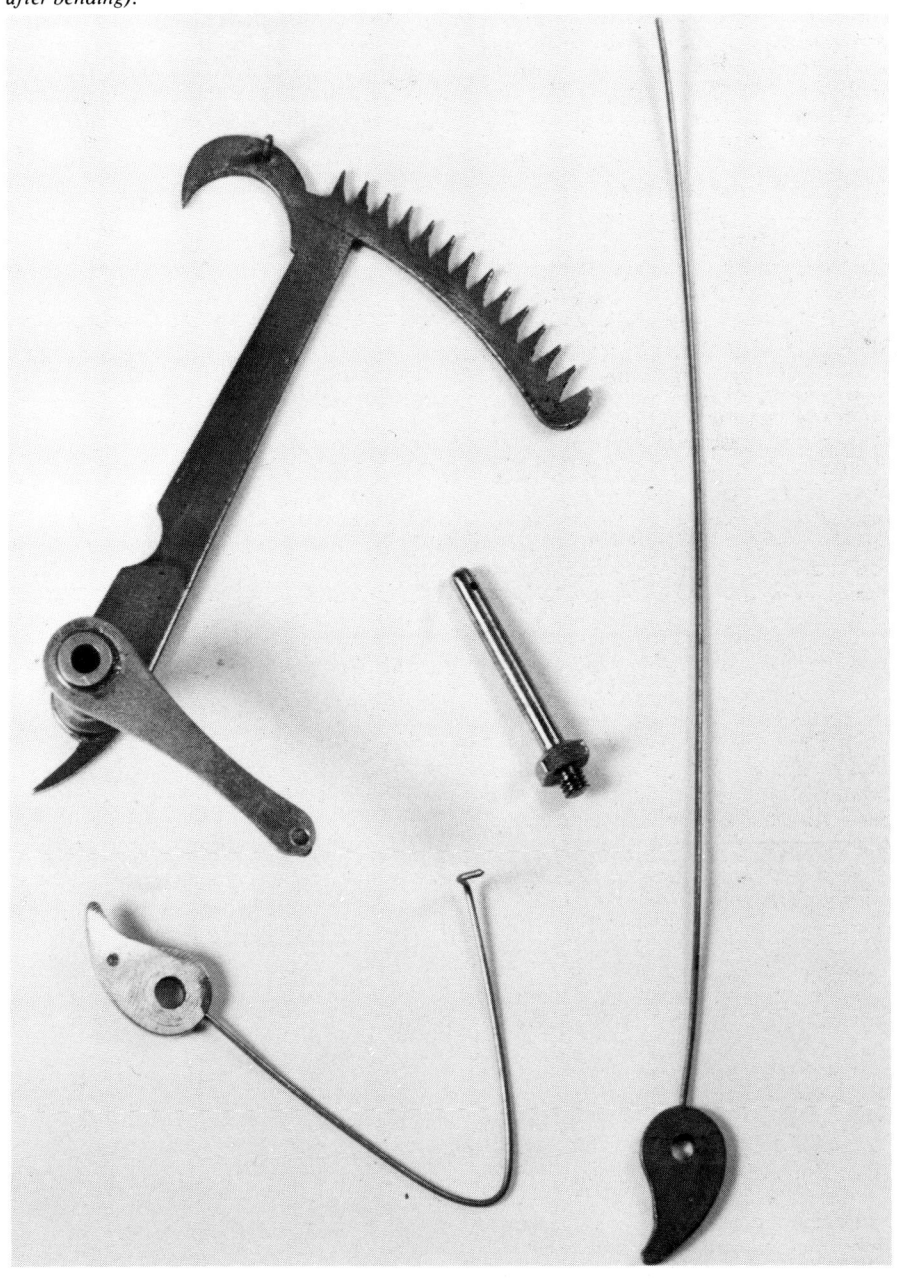

**THE FLY**

I described the fly arbor earlier, along with the rest of the strike train arbors. The fly itself is shown in the drawing, Fig 44, and is a straightforward component to make.

The two blades of the fly are cut and filed up from a piece of 20 swg brass sheet to the dimensions shown. The fly is not fixed to its arbor, but it is held in place by a simple friction spring as shown, which is riveted to one blade of the fly and locates in the groove machined in the fly arbor. This is done so that when the strike train is arrested, the fly doesn't stop instantly, but revolves on its arbor slightly, thus reducing the load on the pivots, etc. The blades are, therefore, silver soldered to two small tubes as shown, and the arbor is a nice running fit in the tubes. As it can be quite difficult to solder these four parts together accurately in line, I assemble the fly using one length of tube, 1" long. This is easily made by drilling and reaming a length of 3/16" diameter brass rod, using a 3/32" reamer. It is then far easier to solder the parts together, after which you can cut away the central part of the tube which is not needed, and thus guarantee that the two remaining pieces of tube are exactly in line.

I have made a very simple jig to hold the parts of the fly during the soldering operation. This consists of a piece of steel bar, about 1" wide, with a groove 3/16" wide by 3/32" deep milled across its width. The tube then sits in this groove, and the blades of the fly can be firmly clamped in place on the jig exactly centrally to the tube whilst the soldering is completed.

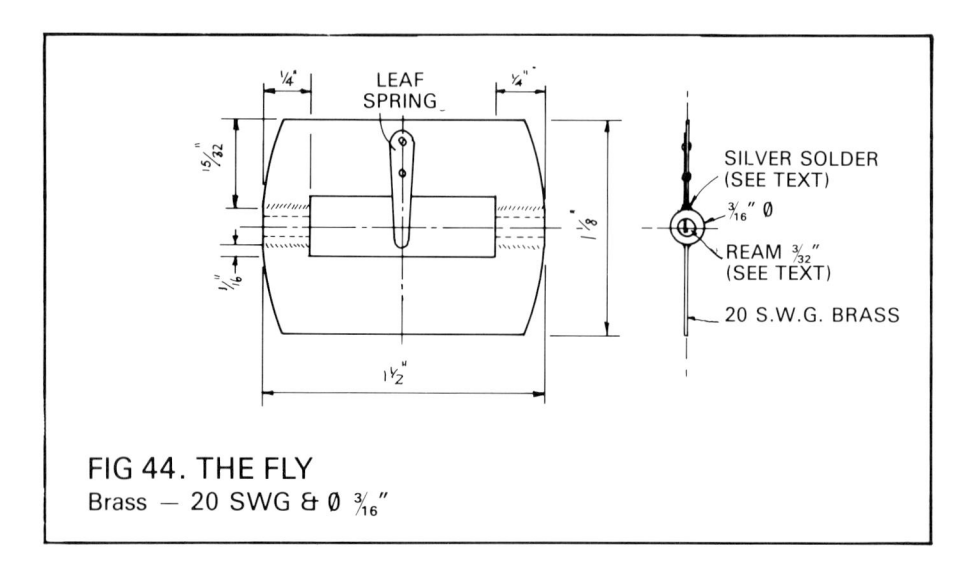

The friction spring is made from either a piece of well hammered 20 swg brass or a piece of thin clock spring. The spring is simply filed to the shape shown in the drawing and then riveted to the fly. I use tapered brass clock pins as rivets. Drill the spring and fly to take one rivet first, rivet the spring in place, and then drill and fit the second rivet. You should now be able to fit the fly on to its arbor by lifting the end of the spring to ride over the thick part of the arbor and, when you push the fly along the arbor, the spring should click down into the recess in the arbor. The fly should now be a fairly stiff fit on its arbor. On no account should it be so loose that the arbor can spin inside the fly when the strike train is running or the train will strike too fast.

**FIG 44. THE FLY**
Brass — 20 SWG & ∅ 3/16"

# Chapter 10

## THE BELL HAMMER AND HAMMER SPRING

Before I describe the remainder of the strikework on the front plate, and the setting-up process, I will describe the bell hammer and spring, as we need these parts to be in place so that we can count the strikes easily.

As you will see from the drawing, the hammer assembly consists of three parts, the arbor, the hammer shaft and the tail. Start by turning the arbor from a piece of mild steel. This is a simple sawing and filing job, but take care to keep all corners nice and sharp. The only dodgy operation on the shaft is drilling the $3/32''$ hole in the $1/8''$ square to take the round section of the shaft. Make sure that you centre punch exactly in the centre of the square, and that the part is held absolutely vertical in the drilling machine vice! With the hole drilled, a $2\frac{1}{2}''$ length of silver steel rod is silver soldered into this part. This rod is a little on the long side, but this allows some latitude for bending the shaft and fitting the hammer head to line up nicely with the bell after this has been positioned. The hammer head is easily filed up from a piece of $1/2'' \times 1/4''$ mild steel to the dimensions given. Do not fit the hammer head to the shaft until the bell is in position, and the shaft has been bent to suit.

The hammer tail is filed from a piece of $1/16''$ thick mild steel or silver steel. Silver steel is better as the end of the tail which engages with the pins on the pin wheel should be hardened to prevent wear. Drill the hole in the tail piece a few thou undersize, and open it out with a broach or taper reamer until the tail is a very tight push fit onto the arbor. Do not solder the tail onto the arbor yet, as you have to adjust its angle on the arbor so that it will engage with the pin wheel properly.

The hammer spring is again fabricated from two parts silver soldered together. If you prefer, you can buy a spring from one of the suppliers of replacement clock parts mentioned earlier in the series. These springs are supplied in the rough state, and need thinning a little to obtain the correct strength, and final finishing etc. If you would rather make your own spring, the details are shown in Fig 46. The foot is filed up from a piece of mild steel, $3/16'' \times 3/8'' \times 1\frac{1}{4}''$ long, to the dimensions shown. The spring itself can be made from mild steel, but again is better made from silver steel, hardened and tempered after shaping. Do not bend the hammer end of the spring until the spring has been silver soldered to its foot and fitted to the back plate.

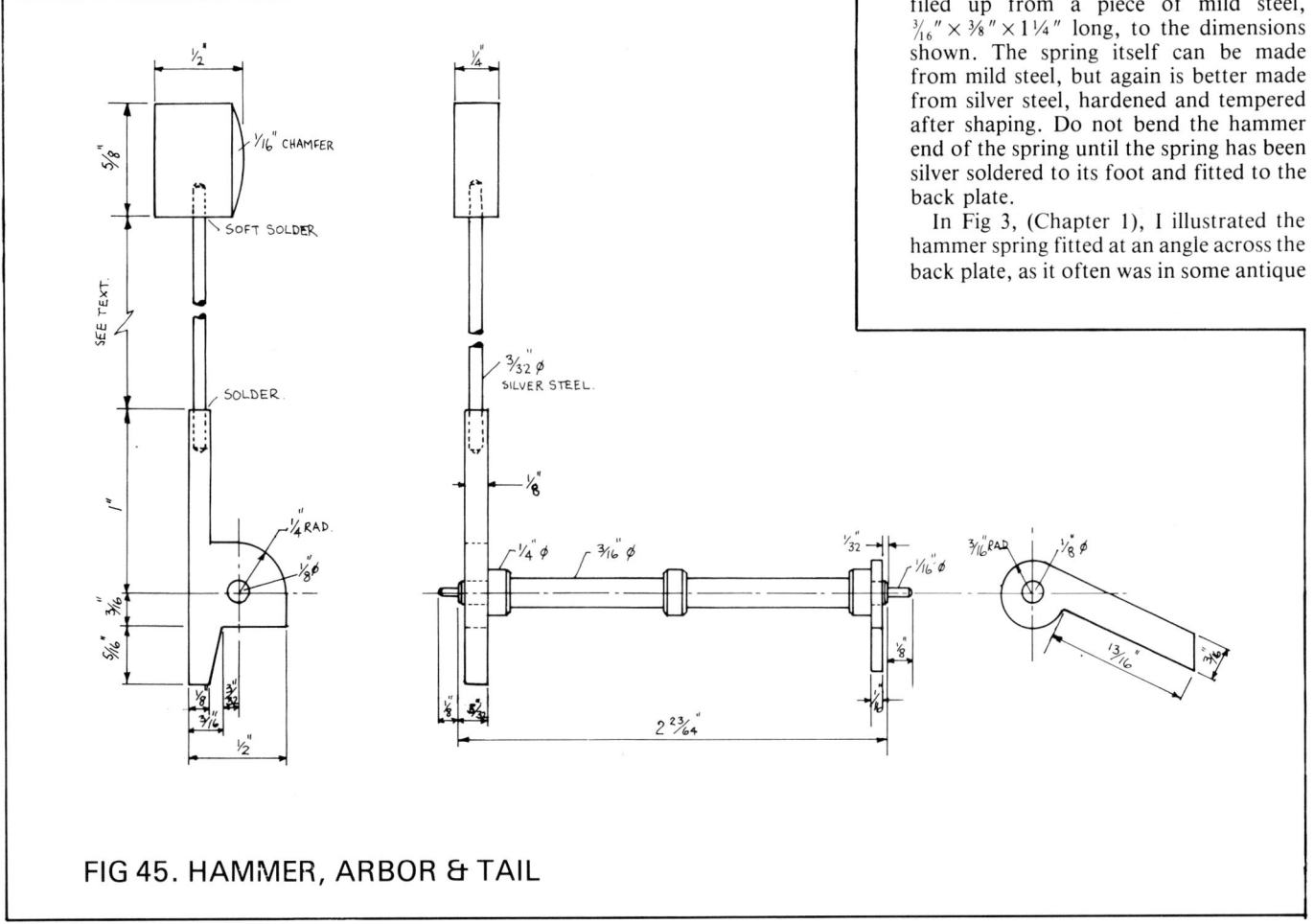

In Fig 3, (Chapter 1), I illustrated the hammer spring fitted at an angle across the back plate, as it often was in some antique clocks. However, I find it far easier to fit the spring parallel to the edge of the plate as shown in the photograph. Drill a clearance hole for the spring fixing screw in the position shown, $1/4''$ in from the left hand edge of the back plate (looking from the front of the plate) and $13/16''$ up from the bottom of the plate. Fit the spring with a screw through from the back of the plate so that the spring is parallel to the edge of the plate. Tighten the screw and drill through both the plate and spring foot $1/16''$ diameter to take a large tapered clock pin as a steady or register pin. This pin should be hammered into the foot of the spring, and the hole in the plate should then be opened out with a broach from in-

**FIG 45. HAMMER, ARBOR & TAIL**

silver steel $1/4''$ diameter, and about $2\frac{3}{4}''$ long. The central boss is merely decorative and its dimensions are not critical. Of course, you should take the usual care with the shoulder lengths, and make sure that the pivots are concentric. The only other critical part of the arbor is the position of the shoulder against which the tail will fit. The tail is lifted by the pins in the pin wheel, and the tail should engage in the middle of each pin. Make sure of the exact position of the shoulder by measuring your movement in case you have strayed from the given dimensions a little.

The hammer shaft is fabricated from three parts. The end which is fixed to the arbor is made from a piece of $1/2'' \times 1/8''$

53

side the plate until the pin is a nice push fit without shake.

The next stage is to fit the hammer arbor between the plates. The pivot holes for this arbor are positioned $3/32"$ in from the left hand edge of the plates, and $3/4"$ above the pin wheel pivot holes, as illustrated. Mark out this position on the front plate, fit the plates together and drill, exactly as you did when planting the trains earlier. Don't forget to drill slightly undersize, and open out the pivot holes using a broach as usual.

Assemble the plates and pillars, and check the end play and fit of the hammer arbor between the plates. If all is well, you can go ahead and bend the end of the spring to suit the hammer. The reason for this bend is that the flat top of the spring formed after bending, acts as a stop on the underside of the hammer shaft, and prevents the hammer shaft from hitting the top pillar during striking. Very often, on old clocks, the makers either fitted a stop plate to the pillar, or simply allowed the hammer shaft to hit it. This, I think, is rather crude. The method used for this clock is a far better solution, although it is a little more difficult to set up. The bend in the spring should be in such a position that the hammer shaft operates without any tight spots as the spring is pushed back, but also without any 'floppiness' during operation. When you get it right,

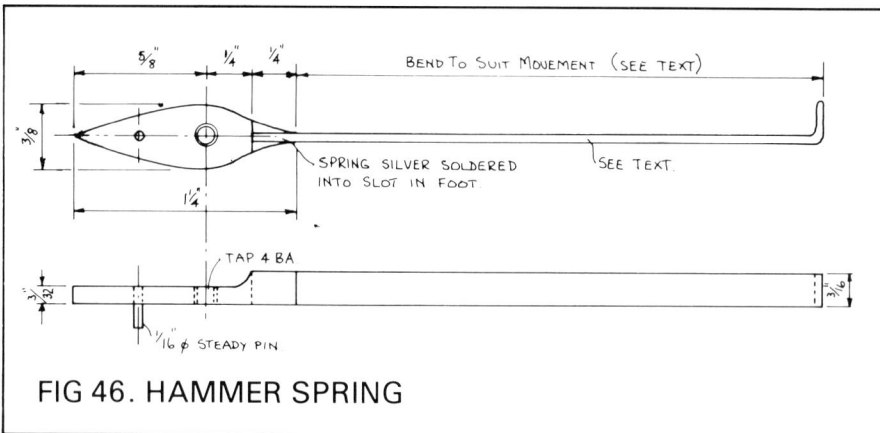

### FIG 46. HAMMER SPRING

the strike train. The spring should be only just strong enough to keep the hammer under control, and not so strong that the strike train has difficulty in starting under its own power when it is released, without using an excessively heavy weight. The spring strength can be adjusted later when the train is assembled by thinning it slightly near its joint with the spring foot to weaken it, or by bending it towards the edge of the plate slightly to strengthen it a little.

With the spring and arbor fitted to the plates, we can now adjust and fix the hammer tail to its arbor. Assemble the plates with the pin wheel, hammer spring and hammer arbor in place. So far, the hammer tail is only pushed onto the arbor, and it should be tight enough to stay in place as the pin wheel is revolved. The tail should slope down slightly as illustrated in Fig 45, and the end of the tail should be in the path of the pins on the wheel. As the wheel revolves clockwise

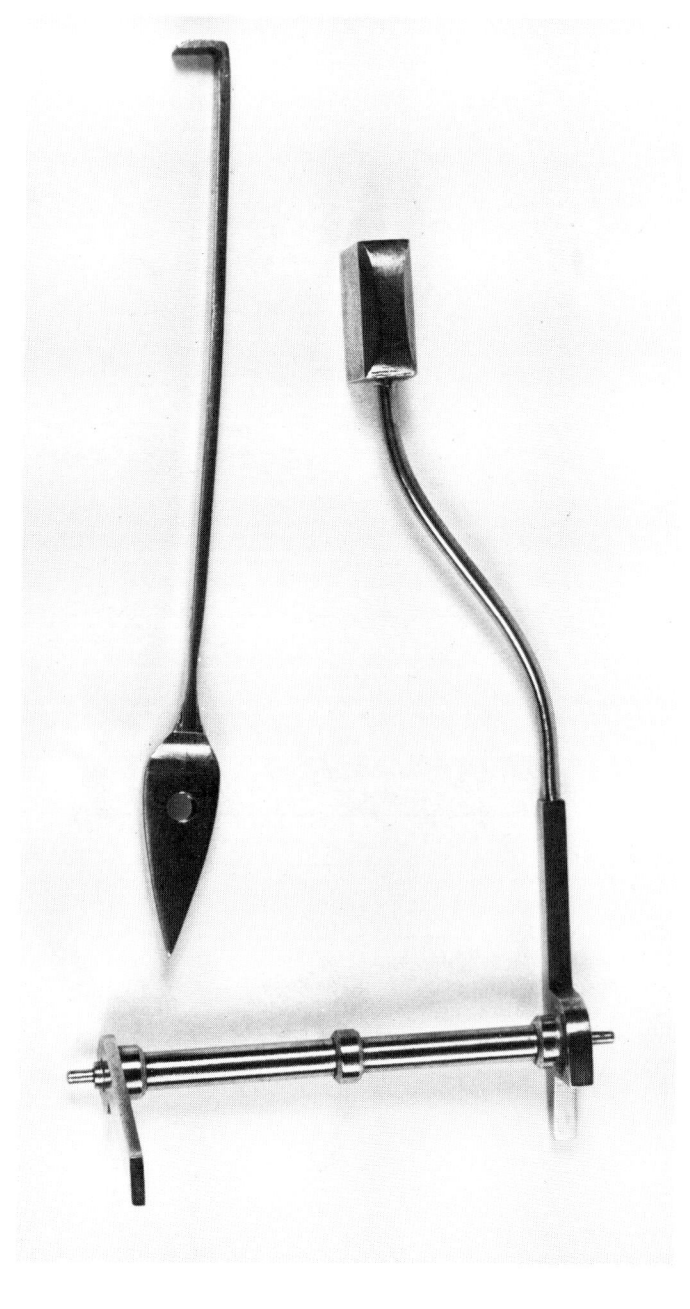

### FIG 47. POSITION OF HAMMER ARBOR & SPRING

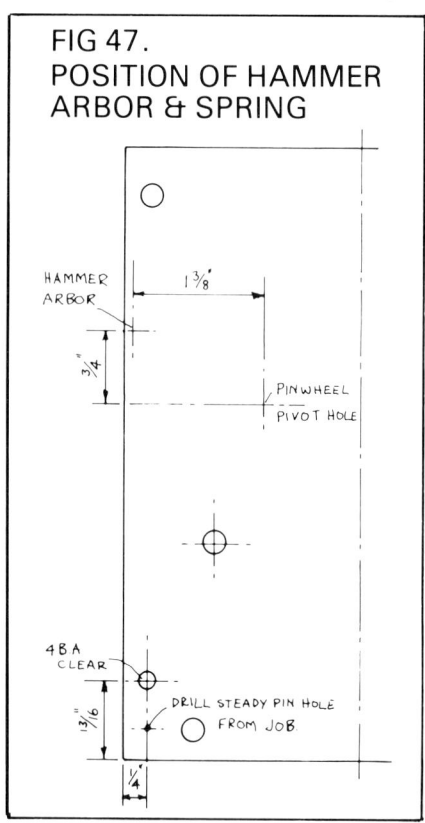

you should be able to pull the hammer shaft back against the pressure of the spring and, on releasing the shaft, it should spring back to its resting place without hitting the pillar, as the underside of the $1/8"$ thick part of the shaft engages with the top of the bent portion of the spring thus stopping any further travel. I have included a photograph of this — Plate 42 — to clarify the position.

The strength of the spring is very difficult to define, and to some extent depends on the smoothness of the rest of

*Plate 40. The Bell Hammer and Spring*

54

(looking from the front), the tail should be lifted by each pin in turn, causing the hammer to be alternately lifted and released. The end of the tail should drop cleanly off each pin and should not hit the next pin on the wheel. If the wheel is revolved slowly by hand, and stopped as soon as the tail drops off a pin, there should be about 1/8" gap between the acting face of the tail and the next pin on the wheel. At maximum lift, the hammer shaft should be at an angle of about 8 degrees to the edge of the plates. The angle of the tail on its arbor should be adjusted in conjunction with the length of the hammer tail to satisfy these conditions. When you have achieved the desired action, take the hammer arbor from the plates taking care not to alter the position of the tail, and soft solder the tail to the arbor. Try the arbor in the plates again to ensure that nothing moved during the soldering operation.

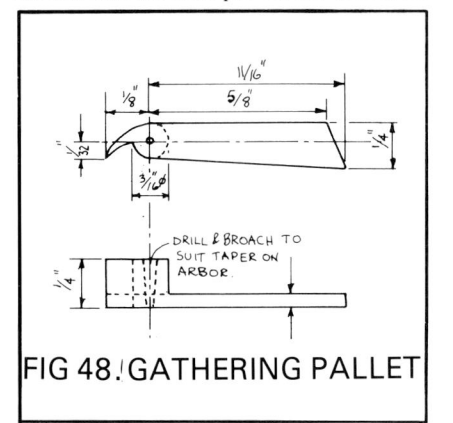

## THE GATHERING PALLET

This is shown in Fig 48. It fits onto the tapered extension of the pallet arbor which projects through the front plate, and engages with the teeth of the rack, driving the rack to the right as the clock strikes. This is another of those components which is fiddly to make, but rough pallets can be bought ready formed, and requiring fitting and finishing. The usual suppliers stock them, and they are very cheap so I usually start with a bought in part. Be careful when ordering a gathering pallet because they come in two types — front and rear locking. We need the front locking type, which means that the tail of the pallet revolves in front of the rack, and the locking pin on the rack projects forwards, thus intercepting the pallet tail as the pallet just finishes gathering the last rack tooth.

If you prefer, you can fabricate the pallet from a piece of 3/8" diameter mild steel rod, drilled centrally in the lathe, and then filed up to form the nib, after which a piece of 1/16" thick steel sheet is silver soldered to the boss just described. Then, the hole in the boss is drilled right through the part and the tail formed by filing. Leave the nib and tail a little on the long side so that they can be filed to fit correctly on final assembly.

## THE LIFTING PIECE AND RACK HOOK

These are shown in Fig 49, again full size to help with marking out the rather complicated shapes. Most of this shaping is purely decorative, and the critical dimensions are given on the drawings.

The collets for both the lifting piece and rack hook are straightforward turning jobs in brass, and should present no problems. They are made in exactly the same way as the wheel collets with the necessary alterations in dimensions.

The tail of the lifting piece is cut from a piece of 20 swg brass sheet using a fine toothed piercing saw. The arm is cut from 1/16" mild steel sheet or strip, and is made up of two parts, silver soldered together. The arm and tail are eventually riveted together on the collet, but this is another case of making the parts a stiff push fit together for adjusting later prior to riveting — so make sure that the tail and the arm are a tight fit onto the collet.

The rearward projection on the lifting piece will eventually pass through a hole in the front plate, to intercept the pin on the warn wheel. This pin will collide with the face of the arm nearest the collet, and the arm must drop away cleanly from the warn pin to allow striking to start. For this reason, the acting face of the arm is filed to a slight curve and is highly polished to reduce friction.

The rack hook is sawn and filed from 1/8" thick mild steel plate to the shape shown. This part should present no problems at this stage, but you may need to do a little bending and fitting on assembly. The collet can be soft soldered to the rack hook as soon as the steelwork is cleaned up and polished.

I should mention here that I always like to see highly finished steelwork on clocks, with nice sharp corners and either a very fine straight grained finish or a high polish. On the best quality work, steelwork and screws are often blued. My method of finishing steelwork, after draw-filing all over to remove any bad marks, is to polish up with progressively finer grades of 'wet and dry' paper, either on emery sticks or wrapped round needle files. To finish flat surfaces, glue strips of various grades of paper to a piece of melamine coated chipboard, and rub the parts face down on this. Gluing the paper to a flat surface stops the paper creasing and helps to prevent rounding the edges of the work. The finest grades of paper will leave a high polish, but you must make sure that all marks are removed with one grade before going on to a finer grade.

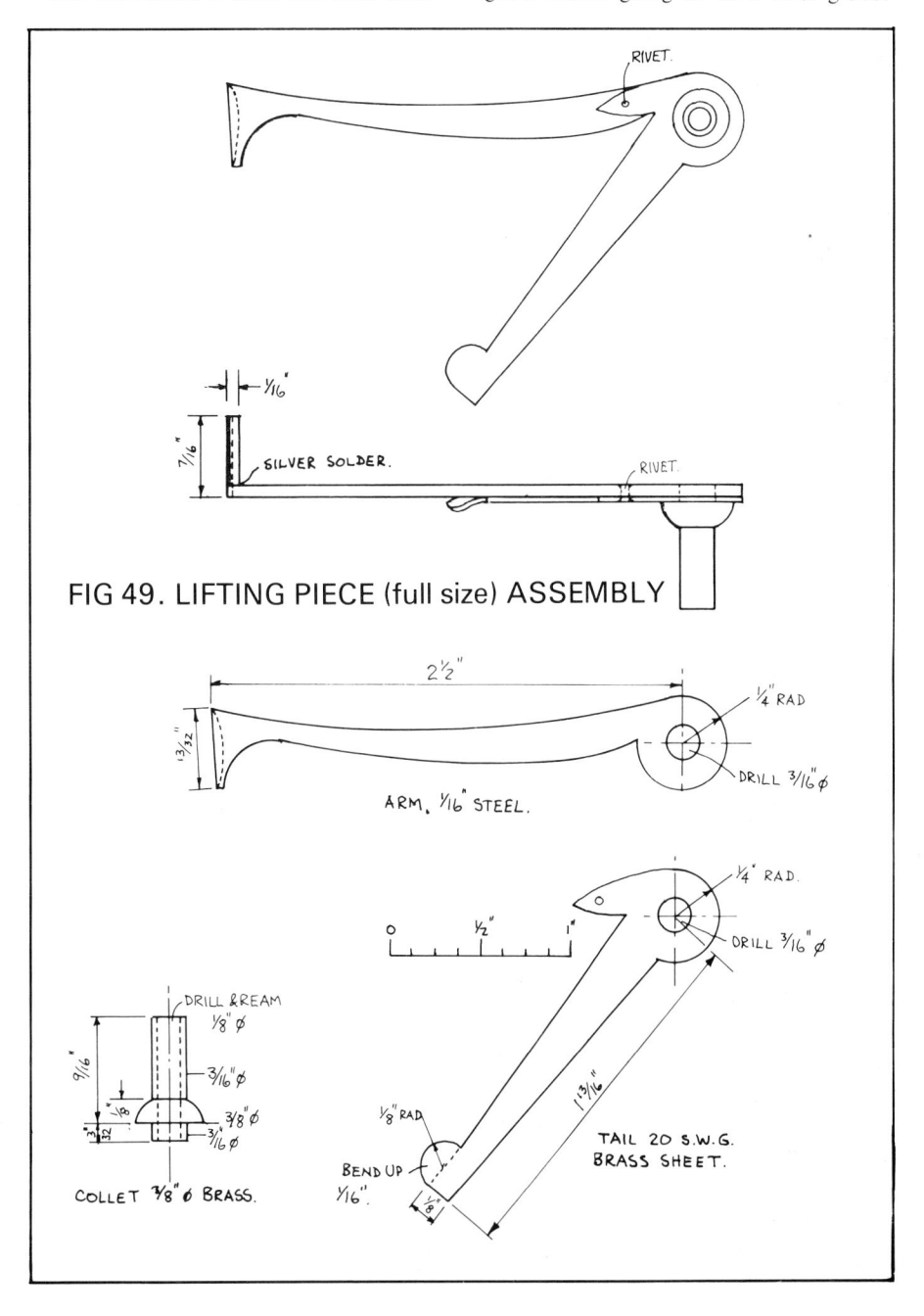

FIG 48. GATHERING PALLET

FIG 49. LIFTING PIECE (full size) ASSEMBLY

*Plate 41. The Rack Hook, the Lifting Piece, the Gathering Pallet and Arbors.*

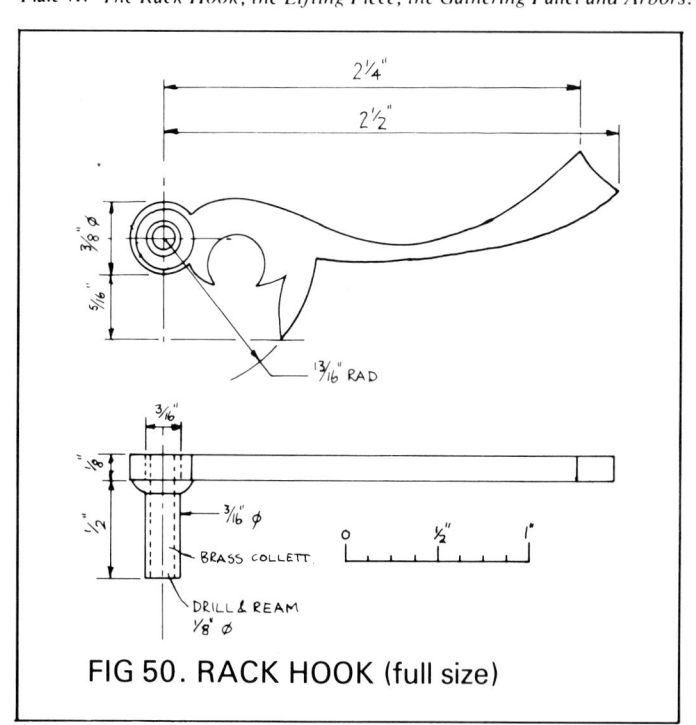

**FIG 50. RACK HOOK (full size)**

The lightest of touches on a soft buff with a little rouge will produce a mirror finish without rounding the edges. I find it much easier to polish the individual parts of a fabricated assembly before soldering the pieces together. It is then a much easier job to clean up and re-polish after assembly.

If you wish to blue any parts, you must bring them to a high polish, and then avoid touching the parts with bare fingers to prevent staining. I find a pair of very thin surgeon's rubber gloves invaluable for this sort of job. After polishing, heat the parts evenly on a gas or electric cooker ring. To do this, bend up a tin-plate tray and cover the bottom of the tray with either fine, CLEAN brass swarf, or a layer of fine dry sand. Bed the part to be blued into this layer so that every bit of the part is in contact with the sand or swarf. Do not bury the part as you need to watch the colour change.

Place the tray over the heat and warm up gently. Have a pair of tweezers and a tray of oil ready to hand. As soon as the part turns a purple-blue colour, lift it from the tray and quench it in the oil. A word of warning — do not use a plastic tray for the oil, as the parts take a few moments to cool and can easily melt the bottom of the tray. This happened to me a few years ago, and the oil was deposited all over the cooker. I still haven't heard the end of this episode, so be warned and keep your wife happy! When the parts are cool, remove from the oil and wipe dry with a clean cloth or tissue.

The final stage with the strike work is to fit the rack and the various levers to the front plate, and to make the bell stand. For those readers who like to have parts ready, we need a bell, so if you are visiting a supplier of clock parts to buy any bits and pieces, you may as well get the bell whilst you are there. Suppliers will also post bells. The item required is a Longcase Clock Bell, either 4" or 4½" in diameter.

*Plate 42. The Bell Hammer and Spring installed in the plates.*

# Chapter 11

**SETTING UP THE STRIKE WORK**

To finish setting up the strike work, we need to fit the rack and its spring, the rack hook and the lifting piece onto the front plate, and to adjust the gathering pallet and the levers to make them work correctly. Before you can do this, you will need to mark out the curved slot in the front plate to the dimensions shown in Fig 51, and cut out the slot with a piercing saw, so that the lifting arm can pass through the plate.

The first stage of setting up is to assemble the strike train between the plates. Until now, we have not been too concerned about the positions of the pins in the pin wheel and the warn wheel, relative to each other. To set up the strike work, it is vital that these wheels are in their correct positions.

When the strike train is at rest, the tail of the hammer arbor must not be resting on a pin in the pin wheel. There must be clearance to allow the train to accelerate and gain momentum when it is released, and before it starts to lift the hammer. At the same time, i.e. when the train is at rest, the pin on the warn wheel must be positioned at about 'a quarter to the hour' when viewed from the front. You must make sure that you get this correct before going on further. It doesn't matter which pin you choose on the pin wheel, because the warn wheel makes one complete revolution per hammer strike. You will probably need to lift the front plate a couple of times to get the relationship correct. I have made it a little easier by fitting the gathering pallet on a round taper on the end of the pallet arbor, thus giving a little latitude for the 'at rest' position. Normally, gathering pallets are fitted on to a square on the arbor, but it is difficult to produce an accurate, tiny square hole in the pallet, and also fitting the pallet on a square further complicates the setting up as there are four 'at rest' positions for the gathering pallet. Fitting the pallet to a round taper allows us to adjust the pallet later.

When you think you have positioned the wheels correctly, pin the front plate on and check again to make sure. As you revolve the train, apply finger pressure to one of the arbors. As the hammer tail drops off each pin in turn, stop the train and make sure that the pin in the warn wheel is in the correct position, and that there is clearance between the hammer tail and the next pin on the wheel.

With the train set up correctly, we can make a start on the rack and lever work. This is probably the most critical part of a striking clock and can be the cause of a lot of trouble later, so it will pay to be particularly careful at this stage. The positions of the centres of motion of the rack and levers are shown in the drawing, Fig 51. These are the positions where the stub arbors, which have already been made, should be screwed into the front plate. If you have made all your components exactly to the dimensions given, your arbors should fit in exactly the same places as on my clock, i.e. to the dimensions given in Fig 51. However, don't count on this! The dimensions are given as a fairly accurate guide, but before committing yourself by drilling holes in the plate, you should set up the parts on false arbors which you can clamp to the plate. You can then check the action of the parts before drilling, and make any adjustments first rather than later!

You will need three false arbors. These are easily made, but they should be made very accurately to avoid problems later. They consist of square pieces of 1/8" brass plate, about 1/2" to 3/4" square, with a 1/8" diameter reamed hole *exactly central* in each piece. Soft solder a 1" length of 1/8" diameter silver steel into each square, making sure that the rods are exactly perpendicular to their square bases. Thus, the brass squares are the same thickness as the squares on the proper arbors, and we can clamp the false arbors to the front plate using small toolmakers clamps. The levers etc., can then be fitted onto the false arbors which can be moved around easily to adjust and obtain the correct action of the levers. When all the levers are working properly, it will then be a simple matter to draw round the brass squares onto the front plate, remove the

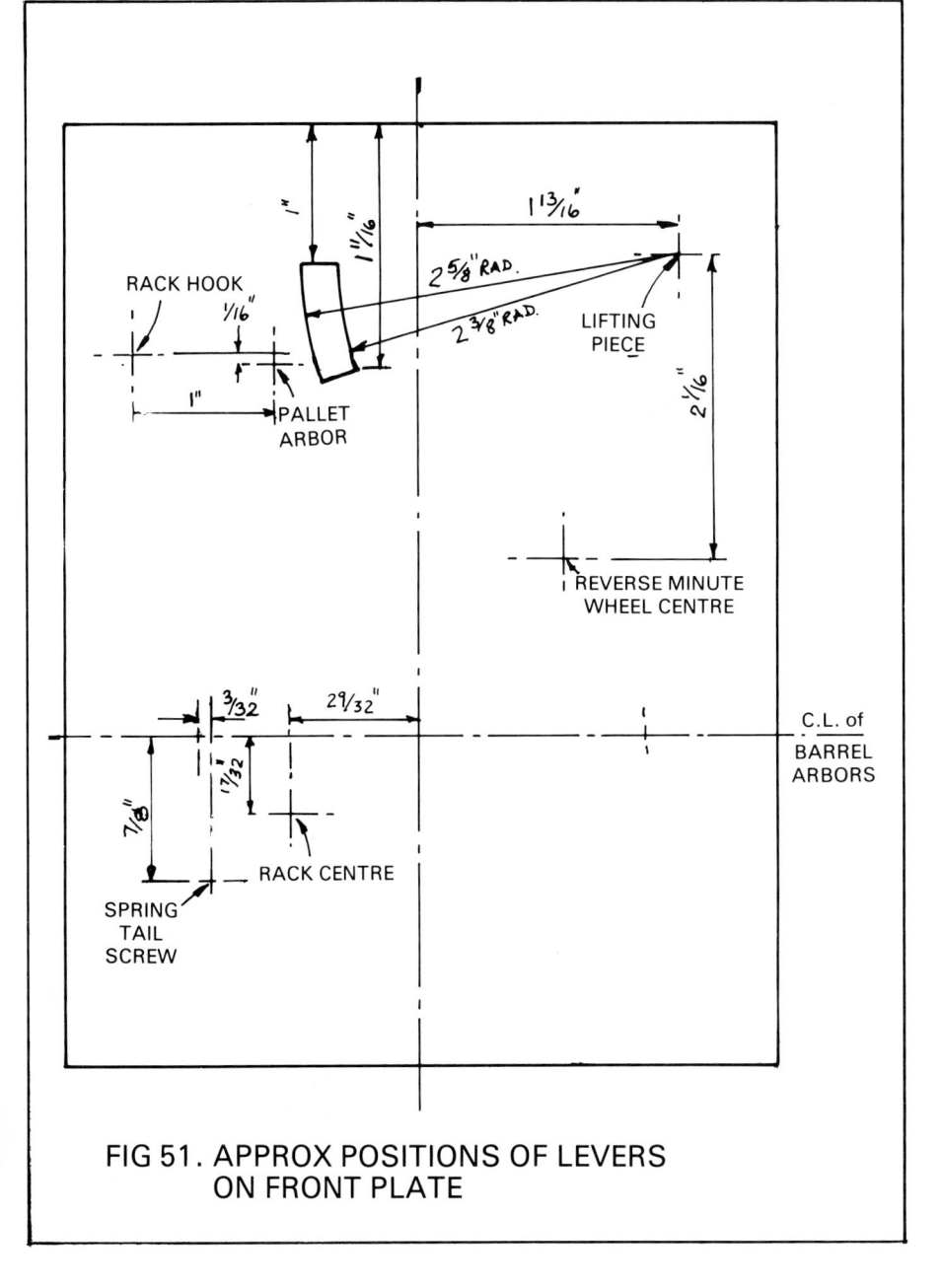

**FIG 51. APPROX POSITIONS OF LEVERS ON FRONT PLATE**

false arbors and draw diagonals across the marked squares to find the centres of the arbors, As the actions of all the levers are inter-related, you must fit them all, and adjust them together before finally drilling the front plate.

Start by positioning the rack and rack hook arbors at the positions indicated. With the rack in position, fit the gathering pallet onto its arbor, with the tail at the front and pointing to the left (see Fig 2),

Make sure that the train is in the 'at rest' position before fitting the pallet. If you give the pallet a light tap, using a hollow punch, it should fit securely on its arbor, and the tail should be just clear of the front face of the rack. At this stage, don't bother to fit the snail in place. We can't fit the rack spring until we know the exact position of the rack itself, so for now, an elastic band (not too strong) hooked round the bottom of the rack and the bottom right hand corner of the front plate will provide the necessary tension. The elastic should pull the top end of the rack anti-clockwise.

Stand the movement up so that the tack hook drops into mesh with the rack teeth and engage the hook with the last but one tooth on the right hand end of the rack. (You will find that you will have to file the small semi-circular depression in the arm of the rack to clear the winding arbor). If you now revolve the train by hand so that the gathering pallet revolves anti-clockwise, the nib of the pallet should engage the last tooth of the rack and pull the rack to the right. I will be most surprised if you get it right first time! You will probably have to file a little at a time off the tip of the pallet to achieve the correct action. The tip of the pallet must not touch the back of the next tooth, and the rack should be pulled to the right by slightly more than one tooth. As this happens, the rack hook should lift up, pass over the tip of the next tooth, drop down into the tooth space and ride up the back of the next tooth slightly. On my movement, the pallet gathers about one and a half teeth. As the pallet passes out of mesh with the tooth, the rack drops back a little until it is arrested by the hook. Each time a tooth is gathered, the train should strike once.

Under no circumstances should you file the rack teeth to try to solve any meshing problems. If there is any problem with the tip of the pallet colliding with the tips of the rack teeth, the position of the teeth relative to the pallet can be adjusted by moving the position of the rack hook slightly.

When you have got the rack operating properly, you can put the snail in place on the hour bridge to adjust the rack tail to strike the correct number at each step of the snail. You will remember that we made the rack tail a tight push fit on its collet to allow for adjustment.

With the snail fitted and the rack back on its temporary arbor, engage the rack hook in the last but one tooth at the right hand end of the rack. Now position the snail, and adjust the angle of the rack tail so that the pin on the tail engages the lowest step on the snail. If you now drive the strike train by hand, it should strike twelve. Revolve the snail so that the rack

*Plate 43. The Front Plate showing the motion and lever work, with the optional datework fitted. The strike train has warned and is about to be released.*

tail engages the highest step, taking care not to move the tail in relation to the rack, and run the train again. It should now strike one.

If the train strikes correctly at both these settings of the snail, and you have made all steps of the snail and all the rack teeth equal pitch as described, the train should strike correctly at every step. Try each step, taking great care not to move the position of the tail, to see if this is so.

If all but one or two steps strike correctly, it is likely that there is some slight difference in the snail steps or possibly tooth pitch, so these should be carefully checked.

If, however, striking is correct at one, two and three, but then gets progressively further out of sequence, it is probably a fault in positioning the pin in the rack tail. If this is the case, refer back to Fig 42 in Chapter 9, which shows the rack/tail geometry and re-read the description carefully. You will find that there is a good deal of 'fiddling about' to be done to get the angle of the rack tail exactly right. You may even find that it is worth making up a trial rack tail with an adjustable pin so that you can experiment with the geometry to make it all work. All of this depends on how closely you have followed the dimensions given. If you have worked exactly to the drawings, all that you will need to do is to adjust the angle of the rack rail. If that doesn't work, go back and re-check all of the dimensions.

When you have got the set up adjusted properly, carefully remove the rack and soft solder the tail in position. After soldering, again re-check the complete strike sequence to make sure that the tail hasn't moved.

As the rack tail drops onto each step of the snail, the nib of the rack hook should drop exactly into one of the teeth of the rack. If this isn't the case, try moving the position of the rack hook pivot. The hook should not drop into a tooth and rest half way up the slope of the tooth back, but it should seat fully home.

At the end of each strike sequence, the train is arrested when the tail of the gathering pallet collides with a steel pin projecting from the front of the rack (see Fig 41, and Plate 39). Now that you have the rack and hook properly positioned, you can file the rack tail to its finished size, and fit the pin to the rack. This again must be done accurately, because the pallet tail has to just miss the pin on its last revolution, but collide with it after the last stroke. The pin must be in such a position on the rack that when the pallet tail is arrested, the pin engages it exactly on the bottom corner of the tail. This is so that when the rack is released on warning, it drops away freely, allowing the pallet to revolve. If the pin engages the pallet too deeply, the rack will be held by the pallet, thus preventing the striking.

When you have decided on the position of the pin, centre punch and drill slightly undersize for a tapered clock pin of about $\frac{1}{16}''$ diameter. The pin can then be pushed in from the back of the rack, and driven home firmly with a few hammer taps. This method is very easy and is strong enough for the job. When the pin is in position, any excess can be filed from the back, and the front end can be cut to length with a pair of nippers and then filed smooth.

With the rack and the hook in position, we are left with the lifting piece. This fits onto an arbor at the top right hand side of the front plate. The brass tail rests over the reverse minute wheel, in the path of the pin in that wheel, and the steel arm passes through the curved slot in the front plate and has the rack hook resting on it. Thus, as the reverse minute wheel revolves, it lifts the tail which in turn lifts the rack hook until the rack drops and the gathering pallet is released. By the time this happens, the part of the lifting piece which projects through the plate is in such a position that it is in the path of the pin in the warn wheel. Thus, when the pallet is released, the strike train will run for a moment until the warn pin collides with the lifting piece. This action is called warning, and it is merely setting the strike train to strike the correct number on the hour. As the train warns — usually at about three or four minutes before the hour — the reverse minute wheel continues to revolve.

During this stage, before the train is released to strike, it is important that there is no tendancy for the lifting piece to try to force the strike train backwards, via the warn pin. This can easily be spotted by watching the fly after warning, with power on the strike train. The fly must not move after warning. This is often the cause of clocks stopping just before they strike, as in effect the going train is trying to lift up the strike weight, which is not a very good state of affairs.

If after setting up you have this problem, it can be cured by either filing the acting face which arrests the warn pin, altering the angle of this face, or altering the position of the pivot point.

To get back to the action of the lifting piece, after the train has warned, the reverse minute wheel continues to revolve, and eventually the tail of the lifting piece drops off the pin on the wheel. This lets the piece drop away from the warn pin and frees the train which can now run until the gathering pallet is arrested by the stop in on the rack, ready for the whole sequence to start again.

I apologise for this lengthy description, but it is most important that the process is fully understood before you try to set it up.

To make the complete strike assembly work, fit the reverse minute wheel to its arbor on the front plate, and clamp the false arbor for the lifting piece at the position indicated on the drawing. Lift the rack hook, and place the lifting piece on its arbor, so that the bent end of the arm passes through the slot in the plate. At this stage, make sure that with the arm resting on the bottom of the slot, the warn wheel is free to revolve. If the warn pin hits the arm, file a little off the bottom of the arm where is passes through the plate.

You now have to adjust the angle of the tail of the lifting piece so that as the reverse minute wheel revolves, its pin lifts the arm and thus the rack hook. The arm must be adjusted such that the rack is not released too early, i.e. before the lifting piece is in the path of the warn pin. To achieve this, you may need to bend the arm of the rack hook a little, or possibly file a little off the top edge of the lifting piece where it bears on the underside of the rack hook. You must also ensure that the angle of the lifting piece is not such that the rack hook is lifted too far. Once the hook has released the rack and the warn pin has been arrested, the lifting piece need not do any more lifting. It simply needs to remain still until it drops off the pin in the minute wheel. This can be achieved by filing the part of the lifting piece tail that acts on the lifting pin. Again, a certain amount of trial and error is needed here, but you will soon see what affects what when you begin to set things up.

Before you finally mark out and drill the holes for the arbors, make a thorough check to ensure that everything works correctly. To do this, stand the movement upright on the bench, and apply 'finger power' to the strike train great wheel, first making sure that the train is in the 'at rest' position. Now revolve the reverse minute wheel anti-clockwise very slowly. The pin should slowly lift the lifting piece, and therefore the rack hook. The rack should drop until it is stopped by the snail, and the strike train should run for a moment. As the train runs, there should be no sign of movement of the hammer shaft.

Continue to revolve the reverse minute wheel slowly, still keeping power on the

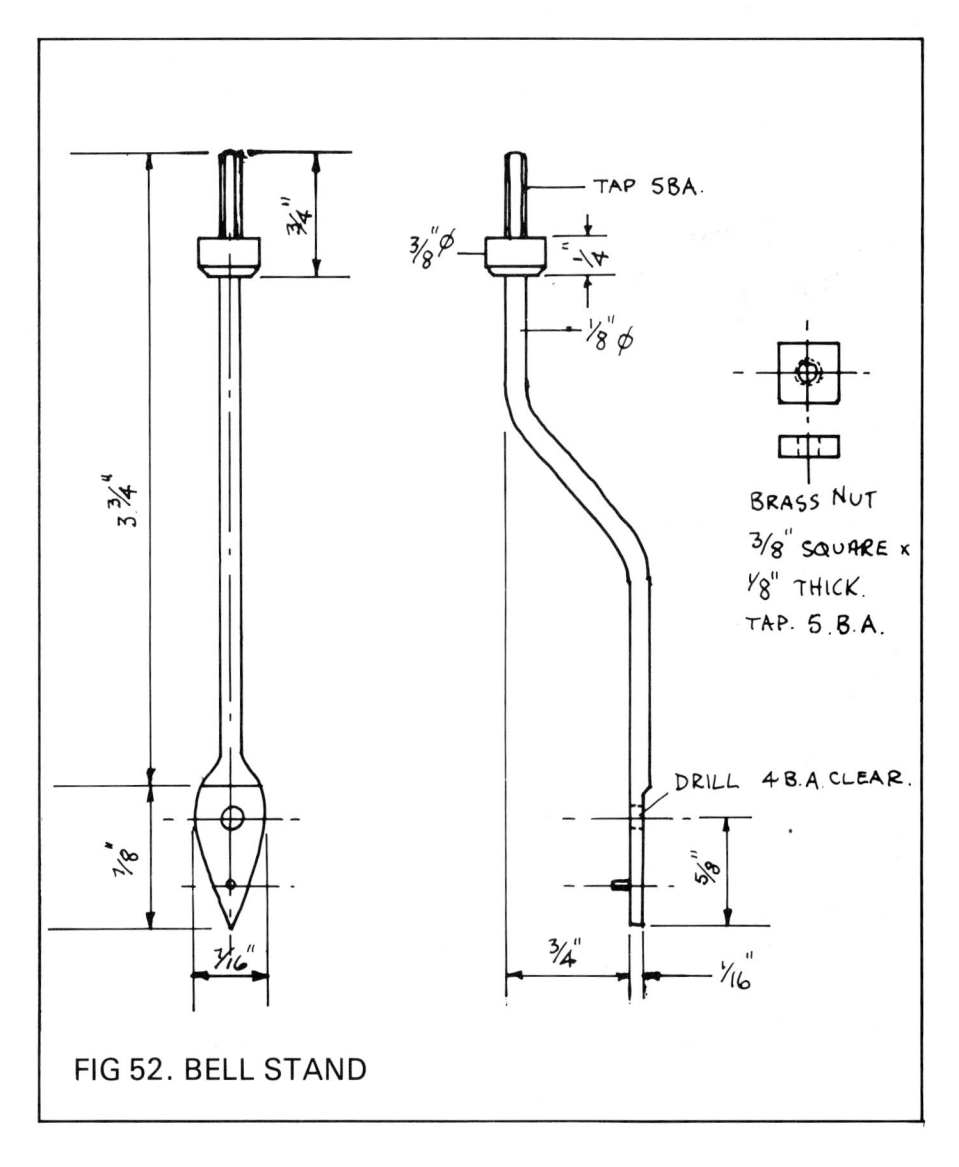

FIG 52. BELL STAND

59

train, and when the lifting piece drops off the pin on the wheel, the strike train should start to run, and strike the correct number as determined by the snail. After the last strike, the gathering pallet should be arrested by the pin on the rack. To be sure, check the complete twelve hour sequence! If it all works properly, well done! You can now solder the lifting piece to its collet and rivet the tail as shown in Fig 49.

When you are happy with the situation, you can fit the permanent stub arbors. Remove all the levers, etc., from the front plate, leaving just the false arbors clamped in place. With a very sharp hard pencil, or very light scriber marks, draw round the four corners of each brass baseplate. Remove the false arbors and draw the diagonals across the corners as marked to find the centre of each arbor. (This is the reason why it is important to make the false arbors accurately). It is now a simple job to drill and tap the plate to take the arbors which have already been made, and these can now be screwed into place.

At the same time, i.e. with the front plate removed, you can drill and tap the screw hole to take the rack spring, after first bending the spring to suit (see Plate 39), clamping the spring foot to the plate to determine its most effective position, and trying its operation. The spring pressure on the rack need not be too strong, as the spring merely assists the rack to drop, and controls any tendancy to bounce on the snail. The actual position of the spring tail is not too important, but as a guide, I have included dimensions to show its position on my movement.

## THE BELL STAND

This is shown in the drawing Fig 52. Construction is self evident. However, I will give a brief description of how to make it, for those who are not too sure. The foot is filed up from a piece of 1/8" thick mild steel, to the dimensions given. The shaft is a piece of 1/8" diameter mild steel rod, 4" long. this is threaded 5 BA for a length of 3/4" at one end. The collar on which the bell will sit, is turned up from a piece of 3/8" diameter mild steel, drilled and tapped 5 BA, which is then firmly screwed onto the threaded part of the rod. Silver solder the rod to the foot, and blend the joint nicely by filing. Finally bend the rod to the shape shown, making any final adjustments when the bell has been fitted.

The Bell Stand is fitted to the backplate as near centrally as possible, so that the rod passes up through the back cock. The mounting hole on my movement is 1 1/8" down from the top edge of the back plate, 5/16" to the right of centre, looking from the back. Again, these dimensions are not critical, as long as the bell is approximately central. With the bell stand screwed on so that it is vertical, drill the usual steady pin hole and fit the pin.

As I have mentioned earlier, you need a longcase Clock bell, about 4" in diameter. This is fixed to the bell stand with the square 5 BA nut illustrated. When you have fitted the bell you can cut the hammer arbor to length and soft solder the hammer head in place. The hammer should strike on the rim of the bell, and the shaft should be bent so that the head is just clear of the bell when it is at rest so that a clear crisp

*Plate 44. The completed clock movement.*

note is sounded without any 'rattles'.

## ASSEMBLY

With all the levers set up, you can now strip and clean the movement ready for testing under power. One point which I forgot to mention earlier, is that all the pivot holes in the plates should have small oil sinks to retain a little oil. These are semi-circular depressions or countersinks on the outside faces of the plates at each pivot hole. To form these, I use the small spherical ended dental drills, which I beg from a dentist friend, after he has finished with them. A few turns of the drill in a wheel brace is quite sufficient, making the outside diameter to the sink just a little under 1/8".

When the movement is stripped, it needs to be thoroughly cleaned, prior to testing. I will leave the description of my polishing methods until later as we need to handle the movement a fair bit yet, whilst fitting the dial, etc. The clock trade use a liquid called 'Horolene' to soak all of the clock parts in to clean them. This is a strong ammonia based concentrate which is diluted with water, and it works wonders, with brass. It is available from the usual suppliers. After soaking in Horolene, a thorough scrub of all parts with a soft brush followed by a rinse in boiling water will be sufficient for now.

With all parts nice and clean, assemble the movement, taking care to assemble the strike train as instructed earlier. You may find that a little dab of 'Loctite' on the gathering pallet is beneficial if your taper is not a very good fit!

The strike train will not work unless the friction spring behind the minute wheel is compressed. Until you have made the hands, put a couple of washers over the end of the centre arbor so that when you put a taper pin through the arbor, the spring is compressed, thus driving the motion work.

Set the movement up on the horse, with temporary weights as already described, set the clock in beat and, with luck, it should all work!

# Chapter 12

**THE DIAL**

It is possible to make all the parts of the dial yourself, on the Myford ML7 lathe, but this means that you will be limited to a Chapter Ring of 10" diameter, as this is the largest diameter that you can swing in the gap of the lathe bed. I now have all my Chapter Rings made by a local engraving specialist, I think that as the dial is the focal point of the clock, it should look really good. It is fairly easy to produce the Roman Numerals on the lathe or milling machine, but Arabic Numerals are a job for the hand engraver! You will see from the photograph of my dial that the minutes are numbered. This is because I wished to make the dial to match the correct style for the period, but this is not absolutely necessary.

The drawing of the Chapter Ring is reproduced full size, 10" diameter, without Arabic Numerals. The type of dial you make depends on how much you wish to get involved with the woodwork for the case. At the Midlands Model Engineering Exhibition, I fortunately had the opportunity of talking to a lot of people who are making this clock, so I was able to sound out what people wanted. It would seem that the vast majority, like myself, are not too keen on complicated cabinetmaking, whilst some are experienced and accomplished cabinetmakers. I have, therefore, decided to describe the construction of a simple, but, I think, good looking case, to take a square dial with a 10" Chapter Ring, so that you will be able to make the whole clock yourself. However, if some of you wish to make a replica of my Clock as shown in the photograph, Plate 47, I will provide dimensioned drawings of the Break Arch case. I am also in the process of arranging for the supply of sets of mouldings, and possibly a full kit of parts to make the case, for those who may prefer to buy the woodwork partly made.

So, at this stage, it is up to you to decide which way you want to go! I have drawn the dial backplate for a 10" Chapter Ring, with an optional Break Arch. If you are going to make a square dial, leave the semi-circular top off the dial plate. If you want to make a larger dial, then increase the proportions accordingly.

For those of you who would rather buy the engraved parts, I have included the addresses of suppliers. Chapter Rings, Seconds Dials and Hands are all available, either engraved or acid etched, and they are all very reasonably priced, and very good quality, so it is well worth enquiring about these if you don't feel up to making them.

So, decisions hopefully made, let's get down to work!

On brass dials, all the engraved and decorative parts are either screwed or pinned to a dial plate, and this plate is attached to the front plate of the movement by three feet, which are pinned to the movement in just the same way as the movement pillars.

*Plate 45. A completed Break Arch Dial with a hand engraved Chapter Ring. Note the matted centre and the date aperture*

The dial plate itself is made from $\frac{1}{16}$" thick brass sheet and again, I use Compo Brass for this. The plate should be marked out exactly square, (or with the optional break arch shape), and then cut and filed exactly to the dimensions given. Now find the exact centre of the square end, and in pencil, draw a horizontal and vertical centre line across the plate.

The hour and minute pipes will pass through a clearance hole which we will eventually drill through the centre of the dial plate. We will not drill this yet as we need to locate the Chapter Ring from the centre. For the moment, just centre punch this position. Now mark out the positions of the winding arbor holes and the hole for the seconds arbor. These will be in the positions indicated on the drawing, but check the exact dimensions from the frontplate of your own movement in case there are any slight differences. Drill the three holes A, B and C — but not the centre hole — using a $\frac{1}{16}$" diameter drill, and leave them at that size for now.

The centre area of the dial plate, the part that shows in the middle of the Chapter Ring, is traditionally matted. On the vast majority of reproduction clocks, this matting is done by acid etching through a patterned etch resist, but I have never seen etched matting that even starts to compare with the traditional thing. The reason for this is that dial matting appears to be a forgotten art or a closely guarded secret! Some people say that matting was done by engraving lines in all directions

61

Plate 46.
The oak case.

Plate 47. The completed Break Arched version of the clock in a mahogany case.

62

**FIG 53. DIAL PLATE**
1/16" Compo Brass

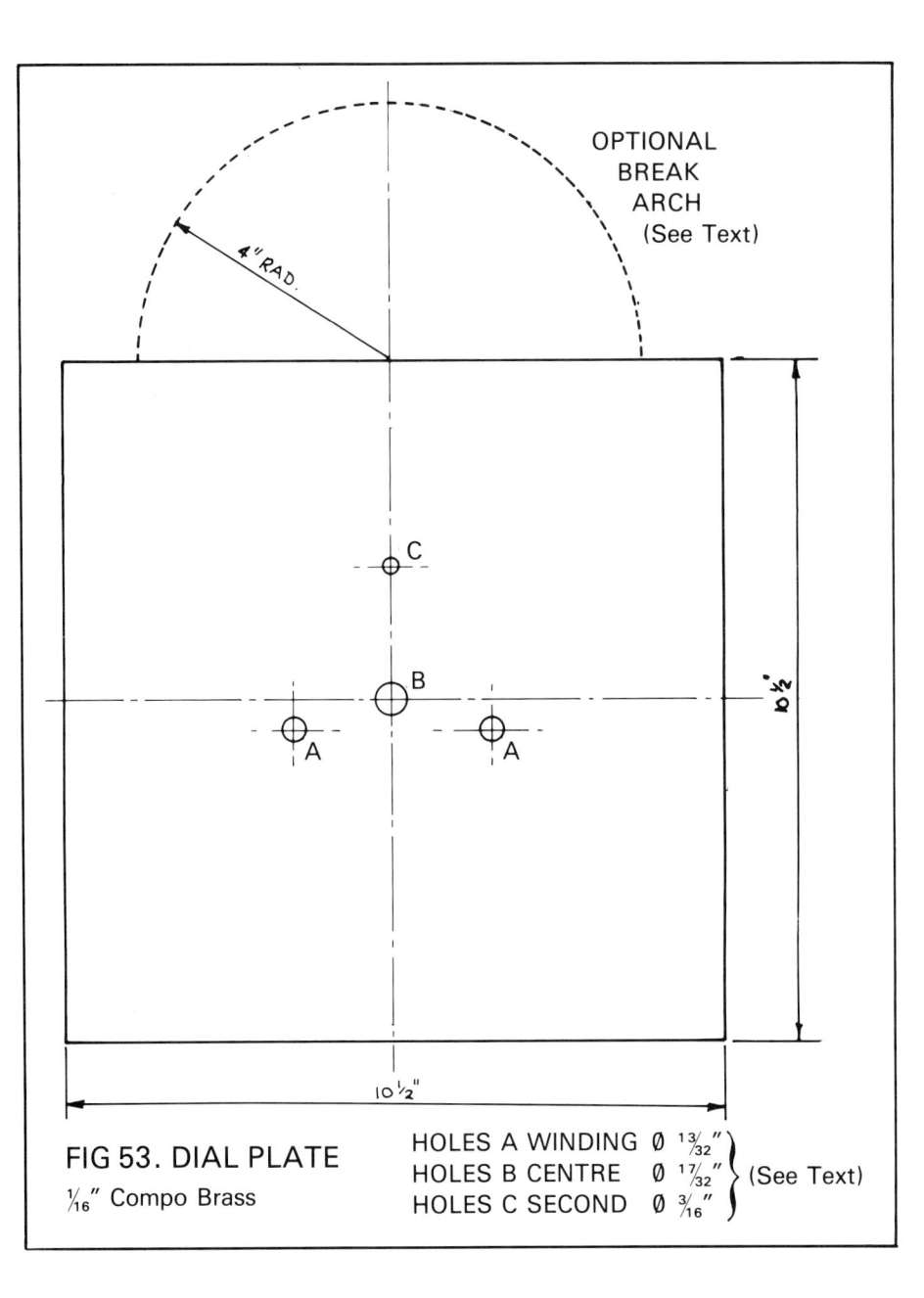

HOLES A WINDING ⌀ 13/32"
HOLES B CENTRE ⌀ 17/32"   } (See Text)
HOLES C SECOND ⌀ 3/16"

Machine the grooves in two sets at right angles to each other, thus producing dozens of tiny square pyramids with sharp points. I make a HSS fly-cutter with a 45 degree point to produce the grooves. This can then be used either on the milling machine if you have one, or on a milling spindle in the lathe, with the work held in the chuck. After milling the grooves, harden the end of the bar and temper to a light straw colour.

To use the punch, mark out a circle on the dial plate a little larger than the inside diameter of the Chapter Ring. Now, support the plate over a block of steel as large as is practical, which you can hold in the vice. Tape a piece of thin card over the surface of the block to prevent marking the back of the dial plate too much. This card also seems to reduce the amount of distortion in the plate. You now have to be prepared for a fairly noisy hour punching the whole of the area inside the circle to produce a nice even matted effect. You will need to experiment a little on a piece of scrap first to obtain an even finish. Take care to keep the punch vertical when you hit it, or you will mark the plate.

As you matt the plate, you will find that it distorts quite a lot. Don't worry too much about this, but do flatten it at intervals and try to keep the distortion to a minimum. As you punch the plate, it work hardens and becomes springy. In the past, I have softened the plate by heating it over a gas ring, but you must do this very carefully to avoid further distortion. Heat it gently and evenly, and then let it cool. Again, try a piece of scrap first to see the effect. I flatten the plate by supporting it on blocks of wood, and pressing the high part down. Careful tapping with a hide mallet with the plate on a flat surface also helps.

There is quite a lot of work involved in matting and flattening a plate. I suspect that this is the reason why a good many people don't bother, and leave the dial over the dial plate, which would be a very tedious operation. The methods available to us all cause distortion of the plate, so we have to flatten it again when matting is completed. I have tried various methods which have been published in the past, including rolling a knurling tool all over the area to be matted, using very heavy pressure. This does work, but is not as effective as the traditional effect. Another method that I have been told is very effective, but which I have not yet had the opportunity to try, involves making a set of hardened knurled silver steel rollers, all the same diameter, which are then mounted on an arbor so that they can revolve freely. The whole lot is then fitted to a handle, so the tool is rather like a paint roller. In use, the tool is not rolled over the plate as you would expect, but it is used to hammer the plate and thus produce the matting. I am told that the final result is very similar to the real thing!

The method which I use is to punch the matting into the surface, using a home made punch. This is made from a piece of 3/4" diameter silver steel, 3" long. The ends are faced and chamfered in the lathe. I then mill 45 degree V grooves at 1/16" pitch across the end face of the bar.

*Plate 48. A hand engraved Chapter Ring and Seconds Ring, filled and silvered.*

centre smooth. However, I think that a well matted dial looks really well, and it certainly is well worth the effort.

When you have finished matting, you can open out the winding holes to $\frac{13}{32}$" diameter, and the seconds arbor hole to $\frac{3}{16}$" diameter. Then put the dial plate to one side until you have made, or bought, a Chapter Ring.

## THE CHAPTER RING

This is made from a 10" diameter piece of Compo Brass, $\frac{1}{16}$" thick. I cut all my brass with a bandsaw, but if you haven't got one, you will have to saw and file the blank by hand. Make the blank a fraction over 10" diameter to allow for cleaning up on the lathe. You now need your lathe faceplate, and a piece of blockboard 10" diameter and $\frac{1}{2}$" thick. Fix the blockboard centrally on the faceplate, using four woodscrews through the slots in the faceplate with large washers under the screw heads. Now, mount the faceplate on the lathe. Mine just fitted at 10" diameter but it was a close thing! If your lathe won't take 10" in the gap, make your dial as large as your lathe will accommodate.

With the faceplate mounted, take a light cut across the edge of the wooden disc to true it up, and take a very light cut across the face of the wood to make sure that it is flat. With a $\frac{7}{64}$" drill in the tailstock, drill the centre of the disc about $\frac{1}{4}$" deep and tap a short length of $\frac{1}{8}$" diameter rod into the hole to act as a central locating peg. Next, find the centre of your brass disc, centre punch it and using dividers mark out the inner edge of the ring, before drilling the centre $\frac{1}{8}$" diameter. Drill four more holes within the circle just marked to take four woodscrews to screw the disc to the wooden faceplate. You can now mount your brass disc on the faceplate ready to start, and give it a coat of marking out blue, which will make it easier to see what you are doing.

The first operation is to true up the outside diameter of the blank. Use a fairly slow speed, and light cuts until the edge just cleans up. From now on, keep the job on the faceplate so that everything will stay concentric.

Grind up a lathe tool to a 60 degree point, and mount it in the tool post pointing towards the headstock. You can now turn the grooves round the ring which enclose the minute and quarter hour marks. These need to be about $\frac{1}{64}$" wide, but this is not critical as long as they are not too wide and clumsy. Use a very slow speed, or even pull the lathe spindle round by hand. You will find that Compo Brass is remarkably easy to cut — this is why it is used for engraving.

When you have cut the circular grooves, mount your dividing equipment on the back end of the lathe mandrel, as described earlier, and set up to obtain 60 divisions. Now turn your 60 degree lathe tool on its side in the tool plast so that it will cut when you wind the cross slide towards you, and pack up the tool so that its point is exactly at centre height. You can now use the lathe as a planing machine by locking the saddle to the bed, feeding the tool in with the top slide, and cutting by winding the cross slide outwards. Thus you can easily 'engrave' all the minute divisions using the 60 division set-up and cutting each stroke between the two circles already machined. Make the cuts the same width as the circles, and do not forget to withdraw the tool before indexing round for the next cut. This may sound basic, but it is remarkably easy to forget. Use the index divisions on the top slide collar to make each cut the same depth. Every fifth division should be made wider, so after you have been all the way round, change the tool for another, again ground to 60 degrees, but with a flat tip about $\frac{1}{16}$" wide and re-cut every fifth mark to the same depth as before. Do not grind a flat on the first tool as you will need it again later.

When you have finished the minutes, set up for 48 divisions, and put the sharp pointed tool back in place. Line up the work so that the tool tip is exactly central in one of your wide minute divisions, then wind the top slide in ready to cut the quarter hour marks. Cut these from inside the circle marked on the disc until the tool runs into the groove already machined. Every fourth line should coincide exactly with every fifth minute division, and again, after you have been all the way round, change the tool and widen the divisions which line up with the wide minute strokes. The quarter hour divisions are entirely optional, and so if this all sounds too complicated you can leave them off.

When all the divisions have been cut, drill a small dimple in the central locating pin using your smallest centre drill so that you can locate the point of your dividers exactly centrally. You can then carefully unscrew the faceplate from the lathe to mark out the numerals.

To make this job easier, I have made a full size drawing of the chapter ring, which you should trace as accurately as possible. The top and bottom serifs of the numerals can be marked out easily with dividers from the centre of the plate. When this is done, lay the tracing over the job so that the 60, 15, 30 and 45 minute divisions are accurately aligned with four of your wide minute marks and tape the tracing to the job with adhesive tape. You can now carefully prick through the tracing with a very sharp scriber to give you enough reference points to accurately scribe the outline of each numeral using a sharp scriber and a steel rule. Note that the 'I' strokes are not always radial lines! The decorative motifs between each numeral are also optional and these can be omitted if you prefer.

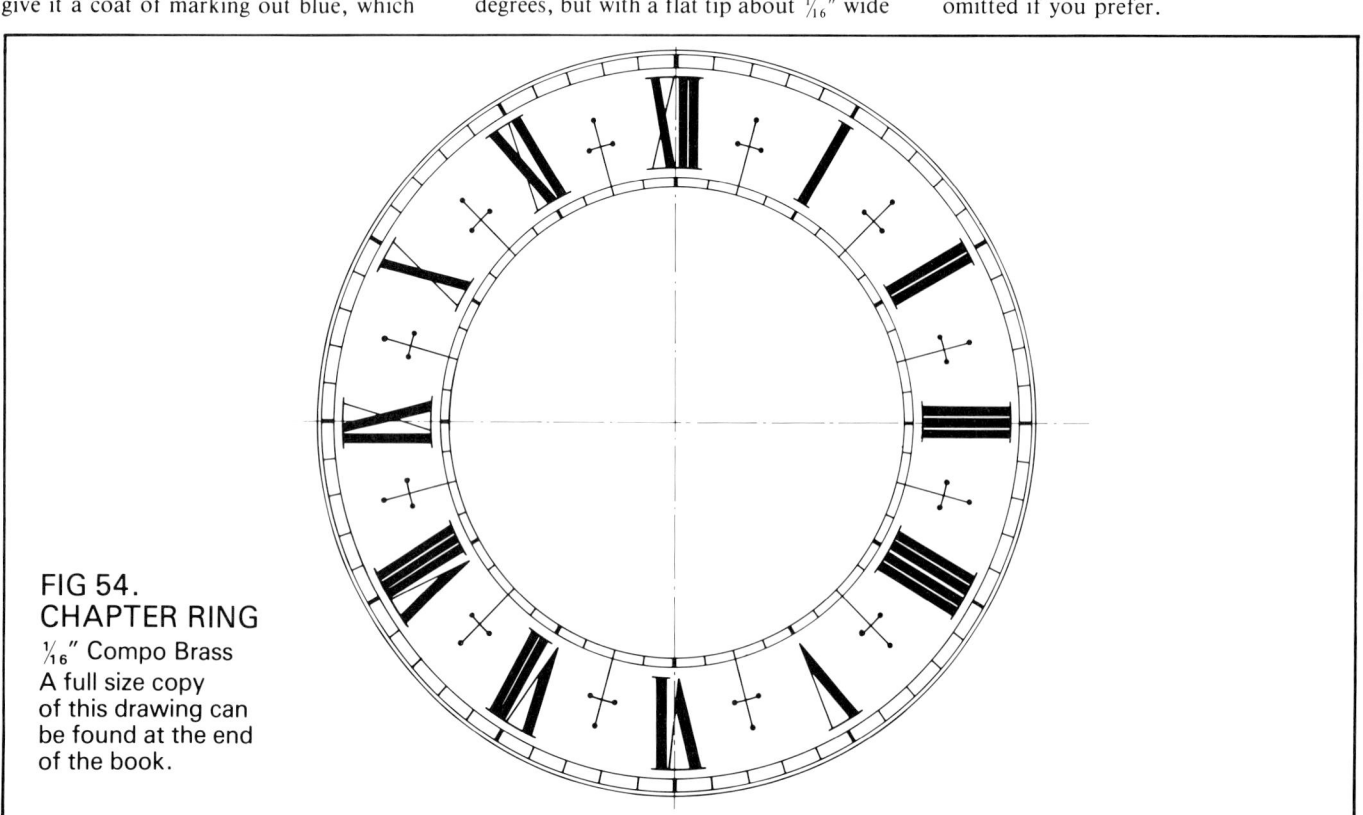

FIG 54.
CHAPTER RING
$\frac{1}{16}$" Compo Brass
A full size copy of this drawing can be found at the end of the book.

If you have decided to make your own Chapter Ring, you should now have a 10" diameter disc screwed to the faceplate as described, with minute divisions completed, and the numerals marked out from the drawing, ready for 'engraving'.

## ENGRAVING THE NUMERALS

There are two basic methods available to the model engineer to produce successful numerals. They can be planed out in a similar way to the minute divisions, or they can be milled out, either with a milling spindle on the lathe, or on the vertical milling machine. I will describe both methods so that you can use the one which suits you best.

To produce the numerals by planing, we need to be able to mount the cutting tools on the cross slide, as we did for the minute divisions, but we also need to be able to adjust the tools vertically so we must mount them on a vertical slide. The photograph shows my set up, with a boring tool holder clamped to the slide, and a ¼" diameter HSS tool bit clamped in the holder. This is a good method of supporting the cutting tool because it is rigid, and the tools can be changed easily and accurately.

Two tools are required for this method. The thin strokes are produced with the 60 degree tool used for the minute divisions, and the wide strokes with a similar tool, ground to a 60 degree angle but with the point ground back to leave a flat cutting edge about ⁷⁄₆₄" wide. This will produce strokes about ⅛" wide when fed in to cut the strokes. The actual width of cut is not critical, as long as the numerals are not too thick and clumsy looking. One can find considerable variation in numeral width on antique dials, but I think that nice slender numerals look far better than the heavier thicker ones. These two tools will cut all the strokes easily in two or three passes. The only problem is cutting the points of the V's.

The first stage with the numerals is to cut the serifs. These are the curved lines at the top and bottom of each numeral. To cut these, mount the fine pointed tool normally in the tool post, as you would for turning, and at 90 degrees to the blank on the faceplate. Adjust the cross slide so that the tip of the tool lines up with the bottom serifs you have marked out, and lock the cross slide. To cut a serif, advance the saddle until the tool touches the work, and apply the cut very carefully pulling the faceplate round so that you cut an arc exactly the same length as that scribed on your work. As I said earlier, Compo Brass cuts very easily, so do be very careful not to slip and cut too far. During each cut, it is wise to lock the saddle to the bed to produce a consistent depth of cut. Take two or three cuts until you think the stroke is wide enough, and clamp a piece of metal to the lathe bed to act as a depth stop to ensure a consistent depth of cut all round the ring. If your lathe is fitted with a leadscrew handwheel, you can of course use this to advance the saddle via the leadscrew, using the graduations on the wheel as an indication of the depth of cut. When you have got the depth correct, go ahead and cut all of the serifs. A word of warning — don't forget to withdraw the tool well out of the way when moving on to the next serif. This sounds very basic, but you'll be surprised how easy it is to forget to do this, and thus ruin a lot of work! When you have cut all of the bottom serifs, wind the tool out to line up with the top ones and repeat the operation. If you decide to add the optional decorations between the numerals, you can also cut the crossbars on these whilst the tool is set up.

Cutting the serifs first makes it easier to cut the strokes of the numerals as they provide clearly defined start and finish points for each cut. I find it best to cut all of the thin strokes first. Set up the pointed tool on the vertical slide as illustrated, so that the tool cuts when you wind the cross slide towards you. None of the thin strokes are radial lines, so we have to revolve the faceplate until the stroke we wish to cut is exactly horizontal, and then lock the lathe spindle to prevent rotation. I do this by fitting a changewheel to the back end of the spindle as I have already described when setting up for gearcutting. I have included a photograph of this set up for clarification. The screw-in detent engages the teeth of the changewheel, thus locking the spindle. The best way of using this changewheel method is to loosen the nut which holds the wheel in place while you adjust the position of the faceplate, and tighten the nut again when you have got the position right. This way, you are not limited by the number of teeth on the changewheel, and you can get every stroke exactly horizontal. Those of you who have mounted a dividing head on the back of the headstock will be able to use this to accurately revolve the faceplate, and the dividing head mechanism will, of course, prevent the faceplate from moving during a cut.

When you have the stroke which you wish to cut set horizontally, adjust the height of the tool vertically so that its tip lines up with the stroke as marked out. This adjustment is done with the vertical slide, which should be locked after adjustment to make the set up as rigid as possible. Go round the ring, and cut all the thin strokes, again working to a depth stop on the bed, and making these strokes the same thickness as the serifs. It is also wise to cut the outside of the wide strokes of the V's with the pointed tool, thus forming the point of the V's which you can't do with the wide tool. It is possible to clear all the area of the V's, which the wide tool will not cut, by taking a series of overlapping cuts using the pointed tool. The alternative method of clearing the points of the V's is to use a graver, but this method calls for some considerable skill.

When you have cut all of the thin strokes, mount the wide tool on the vertical slide, again so that it cuts when the cross slide is wound towards you. Now go round the ring again, this time setting the wide strokes of the numerals horizontal before cutting each stroke. On each numeral where there is more than one stroke (2, 3, 4, 7, 8 & 12), remember that the strokes are parallel, not radial. Thus, you set one of these strokes to be horizontal, lock the lathe spindle, cut the first stroke, and then move the tool either up or down to line up with the next stroke of the

*Plate 49. The tool set up on the vertical slide to cut the numerals.*

*Plate 50. The dividing set-up used to lock the mandrel when engraving the numerals.*

these must be removed without rounding over the crisp edges of the engraving. The traditional finish on brass chapter rings is a coarse circular grain. Thus, to remove the burrs, mount the faceplate and ring back in the lathe, and with the lathe running apply the grain with coarse emery cloth glued to a flat block of wood. Carefully examine the surface of the ring for scratches, and make sure that these are removed at this stage of the graining process as they will be difficult to remove later.

With all the scratches removed you can now cut the centre from the ring. This is done by mounting your pointed 60 degree tool in the tool post, facing the faceplate and setting the tip of the tool to the correct radius (ie, to leave the inside diameter of the ring at $6\frac{5}{16}"$). Apply the cut gradually, using the top slide to feed the tool in, pulling the faceplate round by hand. After all the work which has just been described, the last thing we want at this stage is to see the chapter ring bouncing round the workshop! Cut through gradually until there is just a very thin web of metal left, and then very carefully break the ring away from the centre disc. Any burrs on the inside edge can be removed by drawfiling with a very fine half round file.

The chapter ring is fitted to the dial plate by four feet riveted to the ring, passing through holes in the dial plate, and with tapered clock pins holding the assembly together. The feet are shown in the drawing, and are straightforward turning jobs. The cross holes for the pins are drilled in exactly the same way as the pillar holes, so

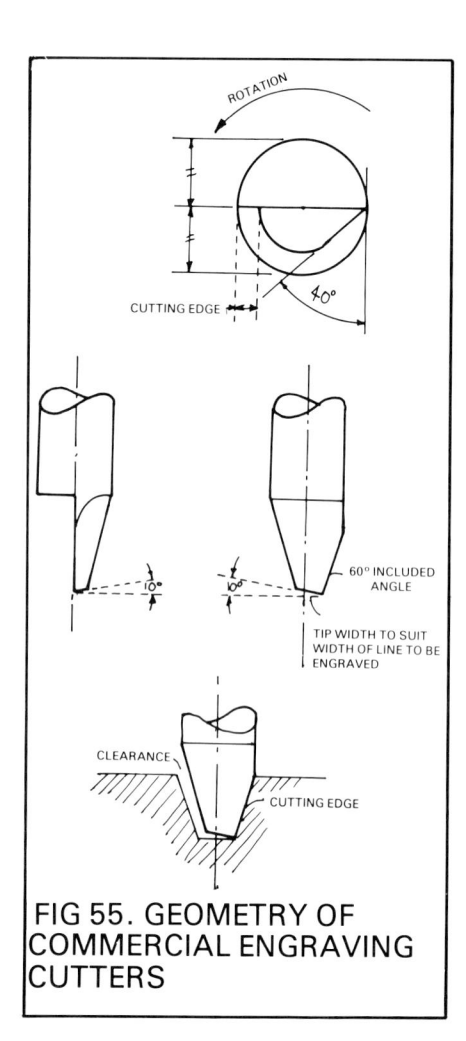

numeral. If you note the readings on the vertical slide collar, you will be able to closely control the spacing of parallel strokes, and make them all exactly the same. Any small variation in spacing of these strokes will be immediately obvious when the chapter ring is filled and silvered, so do take care to space them evenly. As before, set a depth stop on the lathe bed to make all cuts the same depth.

The last stage on the chapter ring, as illustrated, is to drill the small dimples at the ends of the decorative strokes if you have elected to do them. These are simply done with a small drill on the drilling machine, with the depth stop set so that all the dimples are the same depth.

An alternative method of 'engraving' the numerals on the lathe is to mill them, mounting a milling spindle on the vertical slide in place of the planing tools already described, so that a small end mill or D bit can be presented end on to the work. I think that it is easier to cut all of the thin strokes, as already described, by the planing method, and the wide strokes can then be milled. This is because it is difficult to grind a proper tapered engraving cutter with a fine point without having access to a cutter grinding machine. I have included a sketch of an engraving machine cutter as used on pantograph engraving machines. These are very effective, and can be ground to cut different line widths. In commercial machine engraving, it is common practice to cut letters or numerals with thin and thick strokes using one cutter only, in conjunction with multi-line copy. For this method, the cutter is ground to produce the correct width of thin stroke, and the wide strokes are cleared by taking several overlapping cuts. In this way, the corners and points of the wide strokes have only a very tiny radius which is quite acceptable, whereas if the wide strokes were cut with a wide cutter in one pass, the ends of the strokes would be semi-circular, and it would be impossible to produce a pointed stroke.

If you have a cutter grinder, you will be able to make your own cutters. If you don't have one, then plane the thin strokes and mill the wide ones with the smallest diameter end mill that you can obtain so that you keep the corner radii to a minimum. In all other respects, the setting up process for milling the strokes is exactly the same as the planing method.

The final method of 'engraving' is to mill the numerals on a vertical milling machine, and I have included a photograph showing this. I simply clamped the ring, still mounted on its plywood backing, to the machine table, setting each stroke parallel to one axis of the table's travel and milling by normal methods. The comments which I have just made on the cutters for use on the lathe apply equally to the milling machine. Those of you who are making the clock on a Unimat lathe will have to mill all of the numerals. This is quite possible, but to make life easier you will have to make the serifs straight instead of curved.

Whichever method you use to cut the numerals, small burrs will be raised, and

**FIG 55. GEOMETRY OF COMMERCIAL ENGRAVING CUTTERS**

that the pins will pull the ring up tight to the dialplate. The feet are riveted to the ring through $\frac{1}{16}$" diameter holes drilled through the wide strokes at 12, 3, 6 and 9. It is immaterial which of the strokes you choose to drill through.

After drilling the holes in the chapter ring, locate the ring accurately on the dialplate exactly concentric with the centre hole, and with the 60, 15, 30 and 45 minute marks lined up exactly with the vertical and horizontal centre lines on the dialplate. Clamp the ring in position and spot through to mark the positions of the holes for the four feet. You can now rivet the feet in position on the ring, using a flat ended punch to spread the ends of the feet. You can also open out the holes in the dial plate to an easy clearance size to take the feet. Try the ring in position, and insert the pins to make sure that the ring is pulled nicely up to the dialplate.

## FILLING AND SILVERING THE CHAPTER RING

The engraving on the chapter ring is filled with black wax to make it stand out against the silvering which is applied later. The wax used is sold as 'engraving filling wax', and it is rather like black sealing wax as used on letters in days gone by. I have listed suppliers as usual.

To apply the filling, gently heat the ring over a gentle flame on a gas stove, or a low heat on an electric cooker. It is most important that you do not overheat the brass as this will cause the wax to boil, leaving bubbles in the filling. Heat the metal just sufficiently to let the wax flow freely.

Do not try to heat the whole of the ring at the same time. Work round it doing a small area at a time. The idea is to melt a pool of wax to fill a numeral and then, using the edge of a piece of thin card, spread the wax around the hot area to fill all the engraved lines, leaving as little wax as possible on the surface. Any excess wax will take a long time to get off, so use as little as possible. When you have been all round, the chapter ring, the excess wax is removed with wet and dry paper, used wet. To preserve the circular grain, make up a board as shown in the photograph with a handle carrying a block to apply pressure to the abrasive. Using the paper wet, with a little dishwashing liquid in the water, will prevent the paper from clogging up with wax. When all the wax has been removed, wash the ring in warm water (not too hot), and then dry it off. Put it back on the board and re grain it using coarse emery cloth. If you now, very carefully, warm the ring over a gentle heat the wax will melt, forming a very smooth, shiny surface. Be

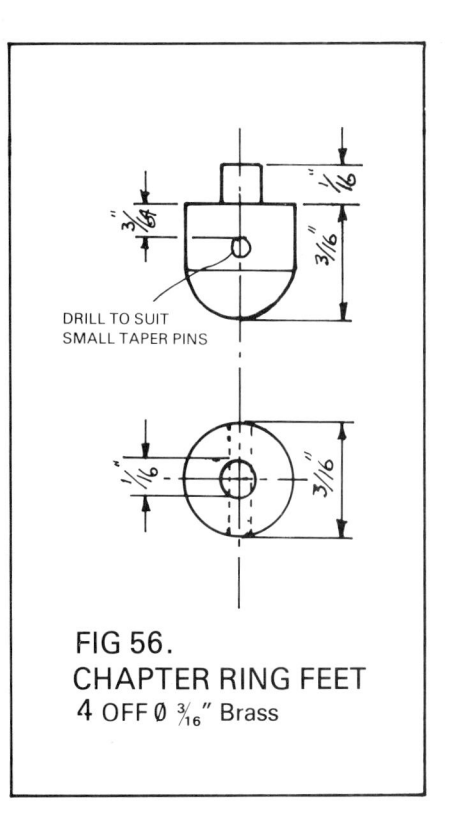

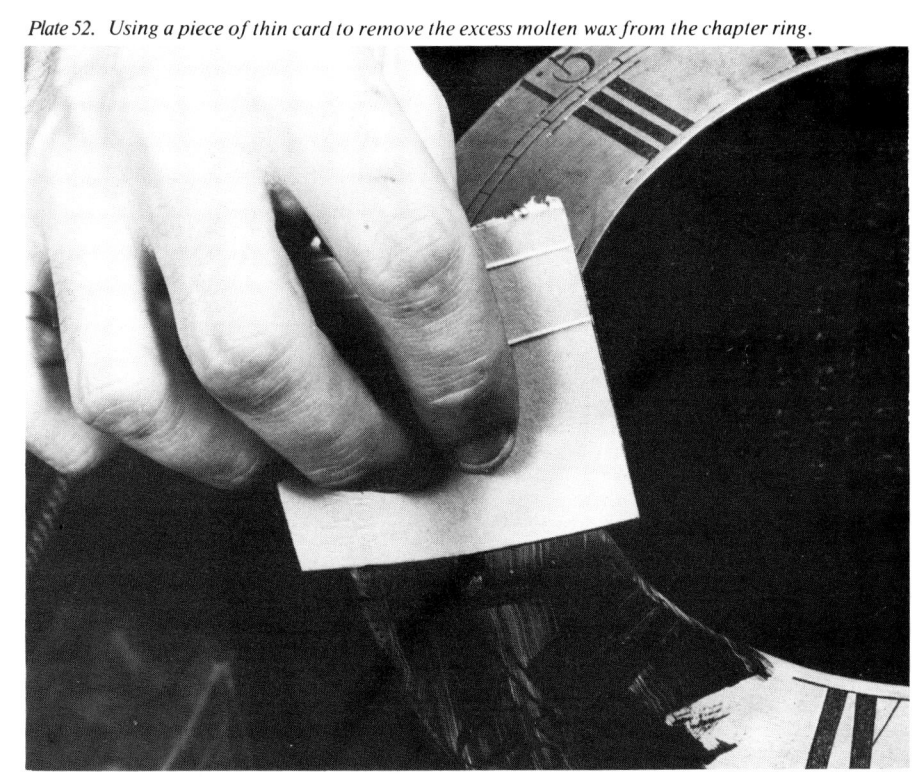

*Plate 51. Milling the numerals on the vertical milling machine.*

*Plate 52. Using a piece of thin card to remove the excess molten wax from the chapter ring.*

**FIG 56.**
**CHAPTER RING FEET**
4 OFF ∅ $\frac{3}{16}$" Brass

extra careful not to overheat the wax at this stage — just sufficient heat to glaze the wax. While you are doing this operation, take care not to touch the clean grained surface as any fingerprints on the brass will produce dark stains when the ring is silvered.

To do the silvering, you will require a jar of silvering powder, (see the list of suppliers) and a packet of cream of tartar, which is available from any chemist. The silvering should be done over the kitchen sink as you need lots of clean cold water. In an old saucer, mix the silvering powder to a thick paste with cold water. In another saucer, make a thick paste of cream of tartar, again with water. Thoroughly wet the chapter ring, and apply the silvering paste with clean cotton wool, keeping up a constant circular motion. Some people sprinkle salt over the dial before they apply the silvering — I don't find that this makes much difference, so stick to the instructions supplied with your powder.

As you rub the silvering compound on, the brass will start to turn white. Keep on applying more paste evenly until the whole surface is a dense, even white colour. When it is, wash the ring under the tap, and then, using a new piece of cotton wool, rub on a good thick coat of the tartar paste, again working evenly in a circular motion. This will cause the surface to turn silver. The effect is quite impressive the first time you do it!

When you have achieved a nice even silver, wash the ring again in clean water, and dry it off with a very clean piece of cloth. Keep your fingers off the surface to avoid stains, and put the ring in a warm place to thoroughly dry off. If, during the silvering process, stains have appeared, you must rub off all the silvering back to a clean grained surface and silver the job again.

The final operation on the chapter ring is to lacquer it to prevent tarnishing. I use Canning's 'Frigilene' Lacquer for this, and apply it with a Humbrol air brush. The lacquer can also be applied successfully with cotton wool, but there is a strong possibility of smearing the wax which is sometimes softened by the lacquer. Apply the lacquer in a dust free place (if there is such a place), and do it in a warm, dry atmosphere to prevent the lacquer 'blooming' and turning an opalescent white colour.

*Plate 53. A simple piece of equipment for hand graining the chapter ring.*

# Chapter 13

## THE SECONDS RING

The Seconds Ring which was shown in Plates 45 and 48 is a hand engraved one by Goodacre Engraving, and is one of their standard items. The ring needed for the dial as described should be 2" diameter by $\frac{1}{16}$" thick. The ring illustrated has the seconds divisions engraved round the outside edge, so the seconds hand must pass over the ring for its tip to register on the divisions. This can create clearance problems between the seconds hand and the back of the hour hand between 11 o'clock and 1 o'clock, as there isn't a lot of space to spare. Many antique clocks used seconds rings with the divisions engraved round the inside edge so that the hand could be recessed into the centre of the seconds dial. Thus, the hand could be kept almost within the thickness of the ring, so solving the clearance problems. I suggest that you use the latter method as it is far less trouble, and it is just as effective.

The problem with making your own seconds ring is that these rings always have Arabic numerals, and there is no easy way to produce these in the average workshop unless you are a good hand engraver, which I certainly am not!

The drawing, Fig 57, shows a seconds ring laid out as suggested. The divisions, (60 in all) are produced in exactly the same way as those on the chapter ring. The only easy way I can think of to do the numerals is to stamp them with $\frac{3}{16}$" high number punches, but the results will not be nearly as nice as engraved numerals. Since both engraved and acid etched rings are readily available quite cheaply, it is probably best to buy one.

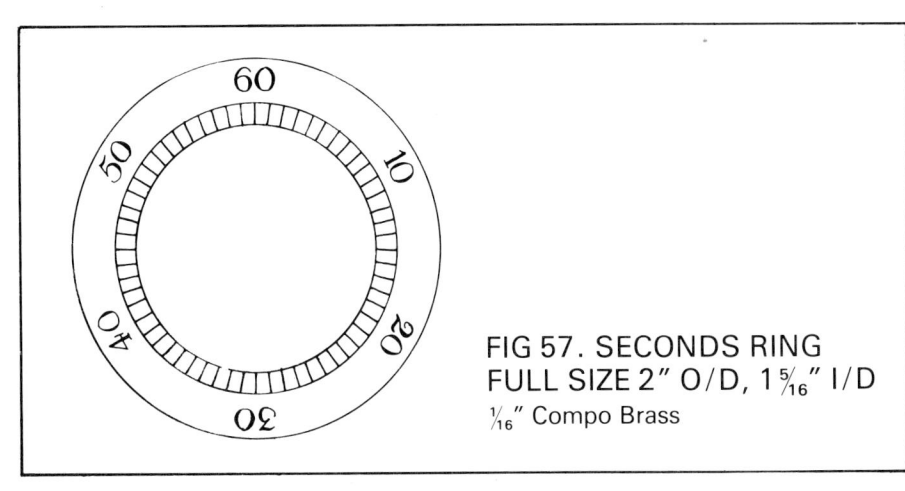

The ring is pinned to the dial plate by two feet similar to those used on the chapter ring. These feet are riveted through holes in the seconds ring before the ring is filled and silvered. Position the feet at the 15 and 45 seconds positions to avoid fouling the date ring if you wish to fit this at a later date. If you take care riveting the feet and clean up the ends of the rivets properly before the seconds ring is grained, you will not be able to see the riveting after the ring is finished.

Finishing the ring is exactly the same procedure as that used for the chapter ring. Apply a good coarse circular grain, fill with black wax and silver, and follow with a coat of lacquer.

## THE DIAL FEET

As I mentioned earlier, the dial plate is fitted to the movement by three feet riveted to the dial and pinned to the front plate of the movement. These are shown in the drawing Fig 58 and are a simple turning job. The $\frac{1}{8}$" diameter spigots at the front end of the feet are riveted into the dial plate, and the $\frac{1}{4}$" spigots pass through holes drilled in the front movement plate and are held in place with the usual taper pins.

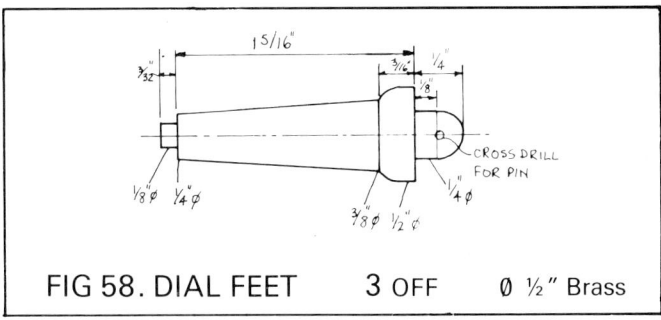

The most important dimensions on these three feet is the shoulder length. On my movement this is $1\frac{5}{16}$" as drawn, but do check this from the job as you may have slight variations. The idea is to position the dial such that there is reasonable clearance between the back of the hour hand and the front face of the chapter ring/seconds hand. The back of the hour hand will of course be on the same plane as the face of the shoulder on the hour pipe, so you can check quite easily to see how long you need to make your dial feet. The taper on the dial feet is purely decorative and thus entirely optional, as there are no clearance problems between the dial and the movement. The operations to make the feet are very similar to those used to make the pillars so there is no need to repeat them here. Do make sure, however, that the shoulder lengths are identical on all three.

The dial feet are positioned on the dial in such a way that the riveted ends of the feet are hidden under the chapter ring. They also need to be placed in such a way that they do not interfere with the parts mounted on the front plate. One foot is positioned at the top of the plate and the other two at the bottom corners of the front plate. To position the feet, remove the front plate from the movement for marking out and drilling. The two bottom feet are positioned $\frac{1}{4}$" in from each side of the plate, and $\frac{1}{4}$" up from its bottom edge. The top foot is $\frac{9}{32}$" down from the top edge and $1\frac{7}{8}$" in from the right hand edge, viewed from the front. Mark out these three positions carefully, centre punch and drill all three holes $\frac{1}{8}$" diameter. Now drill the centre hole of the dial plate $\frac{3}{16}$" diameter, and make up a little peg to fit this hole and the centre arbor hole in the front plate. With this peg, you can locate the dial plate exactly centrally on the dial. Line up the edges of the two plates so that they are all square, and then drill the dial plate holes for the feet $\frac{1}{8}$" diameter, using the front plate as a jig. With this done, you can now open out the holes in the front plate to $\frac{1}{4}$" diameter.

When you rivet the dial feet to the dial, take care to ensure that the holes for the pins are conveniently placed to avoid problems when pinning the dial to the movement. I do this by fixing the feet to the front plate first, inserting the taper pins, then placing the dial in position on the spigots before riveting together in situ. The top foot needs to be positioned so that the taper pin can be inserted from the top and the two bottom feet so that their pins can be inserted at an angle of about 45 degrees downwards from each side so that the ends of the pins are clear of the pillars.

With the feet riveted up and filed flush with the face of the dial plate, you can

**FIG 57. SECONDS RING FULL SIZE 2" O/D, 1 5⁄16" I/D**
$\frac{1}{16}$" Compo Brass

**FIG 58. DIAL FEET    3 OFF    Ø ½" Brass**

remove the dial from the front plate and open out the centre hole to about $^{17}/_{32}$" — a good clearance fit on the hour pipe, which must not touch the dial plate. If you haven't got a drill of this size, drill the hole ½" and open it out with a file.

## DIAL EMBELLISHMENTS

The usual decorative features on brass dials are a cartouche, engraved with the makers name and town, and cast brass spandrels in the corners of the dial plate. On a break arched dial as illustrated, the name is often engraved on a dished boss fitted in the arch with arch spandrels on either side.

The name cartouche is entirely optional, but after all the work done so far, I think it well worth while as it adds that extra little finishing touch. On the square dial as described the cartouche is often an elliptical brass plate pinned to the dial plate just below the centre arbor — an ellipse about 2" wide and 1" high will be about right. On the break arched version, a dished boss 3" in diameter looks right. These are fitted to the dial with the usual foot, pinned to the plate. One central foot will be sufficient as there is no stress involved.

Once again, engraving the names, etc, is a job for the specialist, but fortunately it is quite cheap. The engraver will make the complete cartouche if necessary, or you can make your own blank complete with the foot fitted and send it to the engravers with clear instructions stating exactly what you want engraving. If you choose to do this, it is most important that you make the cartouche from compo (engraving) brass, as other grades of brass are difficult to engrave successfully. It is up to you whether you have your full name, initial and surname or an abbreviation of your first name and full surname. All these were used by the old clockmakers and there is no convention to follow. All my dials are engraved 'ALAN TIMMINS NOTTINGHAM', in the engravers own choice of script and block lettering. They usually add a few 'flourishes' to the tails of T's and A's etc. to make the overall effect quite decorative.

The cast spandrels are again bought in from suppliers, and what a range we have to choose from! Of all the areas of clock restoration, I would think that spandrels have had more attention than anything else. There must be hundreds of styles and sizes available. Nowadays, these are nearly always cast by the lost wax process although there are a few sand castings available. From reputable suppliers the quality is always good. I would advise you to beware of some of the imported articles — especially if they are cheap and you are buying by post — as the quality is not always good and some are made of base metals, or even plastic gilded to look like brass!

The best thing to do, if possible, is to take your dial along to one of the suppliers and try their spandrels on your dial. This way, you will have a choice of several styles that fit, and you can pick the ones you like best. If you have to send for them, get a catalogue first to see the range available.

As for size, the spandrels should fit comfortably in the space available, leaving at least ⅛" at the edge of the dial plate so that the dial will fit easily behind the mask on the hood when you come to fit it to the case. The more decorative spandrels often fit closely to the outer edge of the chapter ring. The more simple 'Cherub' style are smaller, leaving a gap between themselves and the ring. Spandrels are usually measured by the length of their outer edges. On this dial, 3" or 3½" spandrels should fit well, but if you are not able to go and select your own, send a full size sketch of one corner of your dial showing the extent of the chapter ring and ask for spandrels to suit the sketch. This way you will avoid misunderstandings with dimensions. Likewise, if you are making the break arched version, send a sketch of the arch and name boss and ask for arch spandrels to suit.

Spandrels are easily fitted to the dial plate. They are simply screwed from the back of the plate using brass 4 BA screws. When you get your spandrels, they may or may not be drilled and tapped for fixing. If they are not drilled, examine the castings carefully and you will probably find a slight blemish in the surface where the original pattern was drilled, so you may as well drill and tap in the same place. When all the spandrels are drilled, place them on the dial plate and locate them accurately. Then spot through the holes to mark the positions of the clearance holes in the plate. One central screw is sufficient to hold each spandrel. The traditional screws used were square headed and these are available. However, I only use them for restoration work, and I use ordinary cheesehead screws on new dials. If the castings are bent when you get them, flatten them carefully by hand and then bend the corners downwards slightly. If you do this, the spandrel will pull down nice and flat to the plate when you fit the screw.

Some suppliers will sell spandrels ready polished and others sell them as cast. If you get the raw castings it is advisable to spend a little time fettling them using needle files to get rid of any casting flash which is sometimes present. Have a good look at all the holes as well as the edges, especially on sand castings, as these are sometimes almost filled with flash. To polish raw castings, brush them by hand using a brass wire brush. On no account should you buff them with a polishing mop as all this does is to polish the tops of the design and remove cast-in detail. A good vigorous brushing will do no harm and will soon get the castings bright. Finish off with a good hard brushing with a stiff bristle brush. If you rub the brush on a block of chalk (natural chalk is best) you will be surprised at the result. This was the method used in days gone by to polish all brasswork on clocks, and I still use this method as it does no damage and the finish is hard to beat! When you have polished the spandrels, give them a coat of lacquer to prevent tarnishing. The best way to do this is to spray it on with one of the cheap air brushes available from model shops. This way you don't miss all the nooks and crannies in the casting.

Before you finally fit the spandrels to the plate, the plate should be highly polished everywhere outside the chapter ring. The easiest way to do this is to buff it using a soft swansdown mop and rouge, taking care not to round off the edges. After buffing, polish with 'Brasso' or 'Solvol Autosol' chrome cleaner using a soft duster. This process removes the very

*Plate 54. A side view of the clock movement showing the dial feet and also the relative clearances of the hands.*

**FIG 59. HOUR & MINUTE HANDS**
1/16" Mild Steel

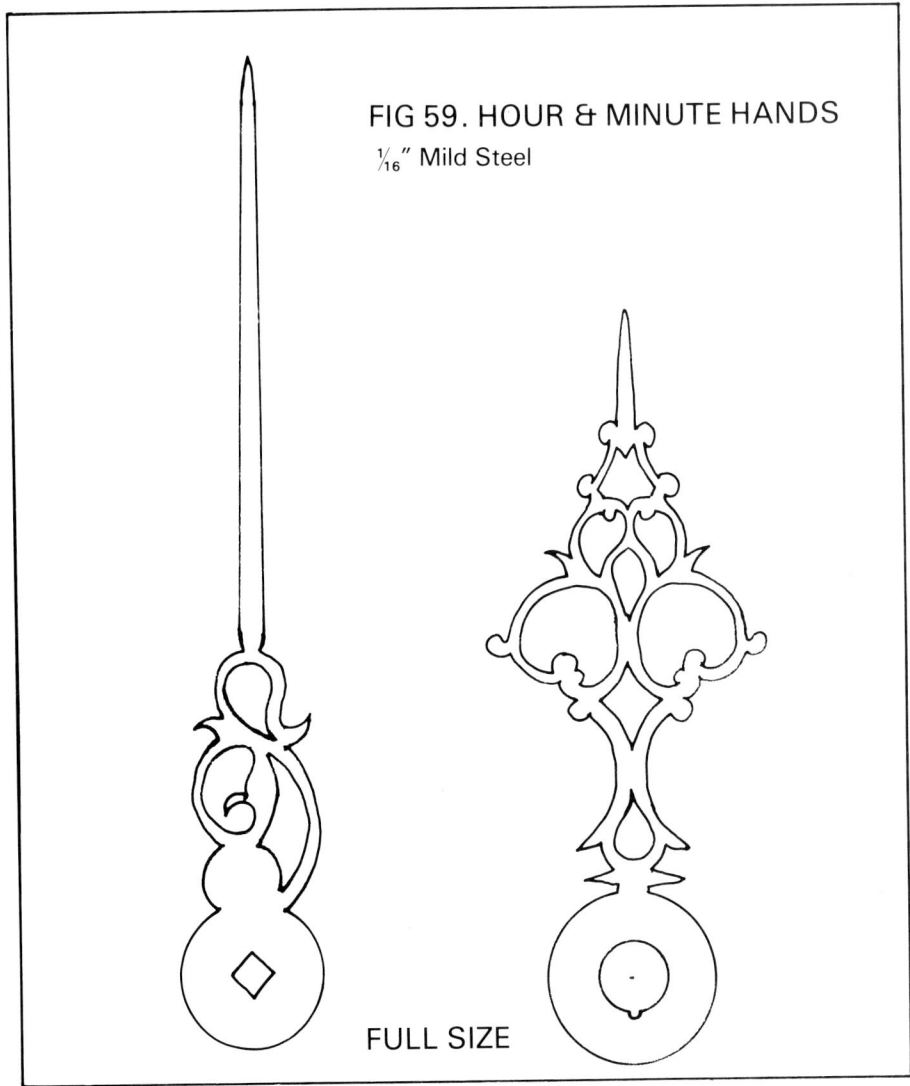

FULL SIZE

fine grain left by the mop. Finally, lacquer the polished area of the plate.

With all the engraved parts, spandrels and dial feet fitted to the dial plate, you can now re-fit the front plate to the movement and fit the dial to make sure that all is well. I usually run the clock with the dial in place but without hands to make sure that there are no problems with pipes rubbing on the dial etc, so you may as well do the same while you are making the hands.

## THE HANDS

These are shown in the photograph, Plate 55, and the drawings Fig 59. On brass dials, the hands are always made of steel, finished by polishing and blueing. To make life easier, the drawings should be printed full size, so I advise you to either copy or trace them and glue the drawings to suitable pieces of steel, ready for cutting out.

Hands are another area of clock restoration that have had a lot of attention from some of the suppliers and again, there are quite a few sizes and styles available ready made, either rough cut or fully finished (apart from centre holes). Obviously, good ready made hands are quite expensive, as they are difficult to make. There is also a vast difference between good and bad samples. This is definitely a case of 'you pays your money and takes your choice!' Alternatively, you grit your teeth and make your own!

The hands illustrated are typical mid 1700's. I should mention here that a very good book showing virtually all dial, hand and case styles is *English Domestic Clocks,* by Cescinsky and Webster. This is available as a reprint, and is a most valuable reference work, well worth a read.

I have suggested that you make the hands from 1/16" thick steel sheet, but 18 swg will do — the thicker material is better. Select good un-marked, flat pieces of steel and clean up one surface with emery cloth. Now glue your drawings of the hands to the steel. I use 'Cow Gum' for this. This is available from artists suppliers and really is the best glue for paper. It has the advantage that when it is set, you can rub off any excess with your finger and it does not reduce the paper to a soggy, pulpy, distorted mess, nor does it cause shrinkage.

I usually cut out all the holes first, finishing with the outside edges. This way, you have plenty of material to get hold of, and the hands are not so liable to bend during sawing. The hands are sawn out using the piercing saw with a fine blade, holding the blanks on the sawing board as you did when crossing out the wheels. I was recently sent a very nice little piercing saw table made by Chronos Designs. This table is fitted on a steel column which can be held in the vice or in a special socket screwed to the bench top. The table itself is made of Beech, with a really useful refinement in that there is an easily adjustable finger clamp to hold the work down during sawing. I always find that holding the work down is more tedious than the actual sawing, so I find this clamp very useful.

With care and practice, you will be able to saw on the lines, leaving very little to clean up. A good set of needle files of various shapes is needed for filing up the hands after sawing. Once again, buy the best quality Swiss files you can afford as the cheaper ones are too coarse and don't last long. I use 'Grobet' files and find them very good.

When you are filing the hands, add a little 'sculptured' effect to the front face by filing vees onto the surface where curves meet producing sharp corners. This does not take long and improves the quality of the finished article. It is also usual to thin the hands towards their tips. You can do this by careful filing followed by rubbing on a sheet of emery cloth on a flat surface.

The hour hand is drilled to be a good close fit on the spigot on its pipe and is held in place with a 10 BA screw passing through a small slot in the hand as in the drawing. The screw passes through this slot into a 10 BA hole drilled and tapped in the shoulder of the pipe. Take care when drilling this hole as there is not a lot of room and the hole must not be too close to the edge of the shoulder or the thread will break through. If the head of the screw interferes with the spigot on the end of the pipe, you can either turn the head down a little, or file a slot in the wall of the spigot to clear the head. This will not be seen when the minute hand is fitted.

The minute hand fits onto the square on its pipe and is held in place by a collet and a taper pin through the end of the centre arbor. Before you file the square hole in the hand, adjust the motion work so that the lifting lever is about to drop off the pin in the reverse minute wheel. Then, holding the reverse minute wheel firmly in position, engage the minute wheel and pipe so that either a corner or a flat on the pipe points up to the twelve position on the dial. You may find that you can't quite line the square up perfectly because of the mesh of the two wheels, so you may have to file the hole in the hand to suit in order to ensure that the strike train is released when the minute hand is exactly on the twelve mark. When you have determined the position of the hole in the hand, drill a hole through the centre of the square a little smaller than the across flats size of your square and open the hole out to a good fit on the pipe using needle files.

The hand collet which holds the minute hand in place against the tension of the friction spring is turned from brass rod. The centre hole in the collet is an easy clearance fit over the end of the centre arbor. The hole drilled across the end of the centre arbor — see Fig 21 on page 28 — should it be in such a position that it is necessary to push the minute hand, pipe and collet back against the friction spring positioned behind the minute wheel before you can insert a small taper pin through the arbor to retain the collet. The easiest way to do this is to drill the cross hole and then make the collet slightly too thick. After a trial assembly, thin the collet until you can push the hand back and insert the pin without too much

*Plate 55. The hands and the hand collet.*

pressure. The hand needs to be held securely enough that it doesn't tend to slip down between 6 and 12 on the dial, and so that the motion work drives effectively. You should also be able to push the minute hand round to adjust the time without using too much force, which may bend the hand.

After fitting the hands correctly, they need blueing. To do this, you must first give them a really good polish. I do this by rubbing on successively finer grades of wet and dry paper, followed by a light buffing on a soft mop with the hand supported on a piece of wood. After polishing, it is most important that you do not touch the surface as finger marks produce stains after blueing.

It is quite difficult to achieve an even purple/blue colour over the hands as the thin parts heat up much more rapidly than the thicker areas. The traditional method is to make a tinplate tray big enough to hold the largest hand and fill this with a layer of either fine, clean brass swarf or dry fine sand. The hand is embedded into the tray so that it is all in contact with the filing, and the whole lot is then gently heated over a gas ring. This provides a nice even heat all over the hand. Heat until the hand turns a nice deep blue and quench immediately in oil. If you go too far and it all turns grey, and this does happen, you will have to re-polish and start again.

Another method of blueing, which is very effective, is to use blueing salts. This is a compound of salts which, when heated, melt and boil at the correct temperature for blueing steel. You simply tie a piece of thin wire to the hand and immerse in the molten salts, removing it from time to time to see how it is going. When it is the right colour, quench in water. If you do use these salts, you must be extremely careful as they melt at a high temperature and can give a very severe burn. There is also a lot of spitting and spluttering when you quench them so again take care. After using the salts, they set into a solid block but can be used many more times. One problem is that the salts are very hygroscopic (absorb vast amounts of water vapour from the atmosphere) and if they get wet, they are useless. Stored in an airtight container they will keep for quite a long time.

The seconds hand is also blued steel, silver soldered to a brass tube or pipe. This pipe is split for part of its length so that it is a tight push fit onto the tapered end of its arbor. You will need to adjust the length of the split so that the hand pushes on as far as possible without rubbing on the dial. Obviously, this hand is polished and blued after soldering to the pipe.

All the hands, by the way, should be of such a length that their tips just overlap the circle containing the relevant divisions. If you haven't used quarter hour divisions, make the hour hand just overlap the bases of the numerals. If you use a seconds ring with the divisions on the inside edge, the hand will be recessed into the centre of the ring so it should be as long as possible, just clearing the inside edge of the ring. When the hands are finished, lacquer them to prevent tarnishing before assembling everything for a test run.

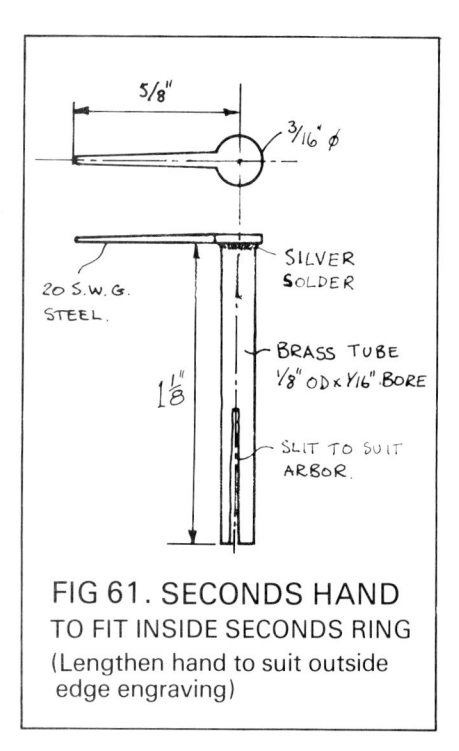

**FIG 60. HAND COLLET**
Ø ⅜" Brass

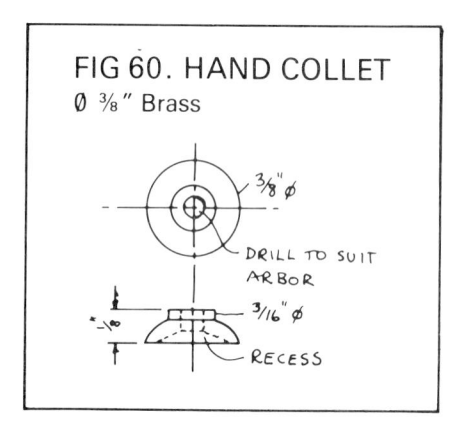

**FIG 61. SECONDS HAND**
TO FIT INSIDE SECONDS RING
(Lengthen hand to suit outside edge engraving)

# Chapter 14

## THE DATEWORK

The datework is an optional extra to the clock, and if you have decided to fit it then the work should be done at this stage, before the final cleaning and assembly of the movement. The method I have chosen to use is, I think, the neatest by far. The date is displayed as an engraved Arabic numeral in a small aperture cut into the bottom half of the matted dial centre. Unlike the usual dial and hand system, this method changes the date in one movement between midnight and about 2 a.m. There are other methods, and these usually alter the date in two stages, mid-day and midnight. The method we shall be using is slightly more difficult to make, but is far superior in the end.

The method of operation is simple. If you look back to Fig 37 on page 46, you will see that the hour pipe has a collar which should have 30 teeth cut on it if you decided then to fit datework, as mentioned in the relevant text. Plates 33 and 37 both show the hour pipe with the teeth cut. This wheel revolves twice a day, and so if we want the mechanism that advances the date to work once a day, we must arrange a 2:1 reduction to the mechanism. This is easily done by mounting another wheel on the front plate, on a stub arbor. This wheel must therefore have 60 teeth to provide the correct ratio.

The extra wheel is mounted on a slightly unusual, long collet to which we solder an arm or 'flag'. Thus, when the clock is running, this flag will revolve once a day, driving the date ring round one step at a time.

The date ring, as you will see from the drawing and photographs, is mounted on rollers on the back of the dial plate. It consists of a brass ring, 6" in diameter, with 31 internally cut teeth. Engraved around the ring are the numbers 1 to 31, lining up with the teeth. As the flag revolves, it comes into mesh with one of the teeth and thus pushes the ring round one step over a period of about two hours. After this time, the flag passes out of mesh with the tooth, leaving the ring with the new date showing through the dial aperture. With this method, the wheel train which drives the date ring is out of mesh with the ring most of the time, thus reducing friction, and also making it possible to change the date at the end of the month by simply revolving the ring by hand when it is not being changed by the clock.

## THE WHEELWORK

By now, the parts needed for the wheelwork should present no problems. The stub arbor is made from ¼" square steel, either turned from the solid or fabricated as described previously. One end is threaded 3 BA to screw into the frontplate. The arbor itself is ⅛" diameter and should, of course, be a free running fit in the collet. The hole for the taper pin should be drilled after the collet has been tried on the arbor. Make sure that there is a little end float.

The collet is unusual in that it is extended above and under the wheel mounting, and it is reamed ⅛" right through to form a pipe to run on the arbor. It is a straightforward item to make, but be sure that the wheel seating and the centre hole are concentric. The flag is simply made by sawing and filing from 1/16" brass sheet. It is best to leave the flag a little over 9/16" long to allow for adjustments later. Similarly, the front face of the flag (marked A on the drawing) should be left to project a little more than indicated, as this face should only just clear the back of the dial and you may have a little more space than I had. The flag is silver soldered to the pipe AFTER the wheel has been soldered into place. The step in the flag is to clear the taper pin used to retain the assembly on its arbor. After a trial assembly, when the acting end of the flag has been filed to length, the end should be radiused and polished as shown in the drawing.

The wheel is 2.240" in diameter, 1/16" thick with a 3/16" centre hole, and is made of 1/16" Compo Brass. It has 60 teeth which are the usual 0.8 Module. There may be some queries regarding the diameter of the 30 tooth wheel on the hour pipe and those amongst you who check on such details will find that this 60 tooth wheel is not the theoretically correct diameter. The reason why these two wheels do not conform to the usual formula is that I wanted a little more space to fit the whole thing together easily, so I made these a little over size — and this makes very little, if any, difference to the meshing of the two wheels. There is no need to cross out this wheel as it revolves very slowly and is not seen behind the dial. After cutting it, clean it up and silver solder it onto

*Plate 56. The datework gearing fitted to the front plate.*

**FIG 62.**

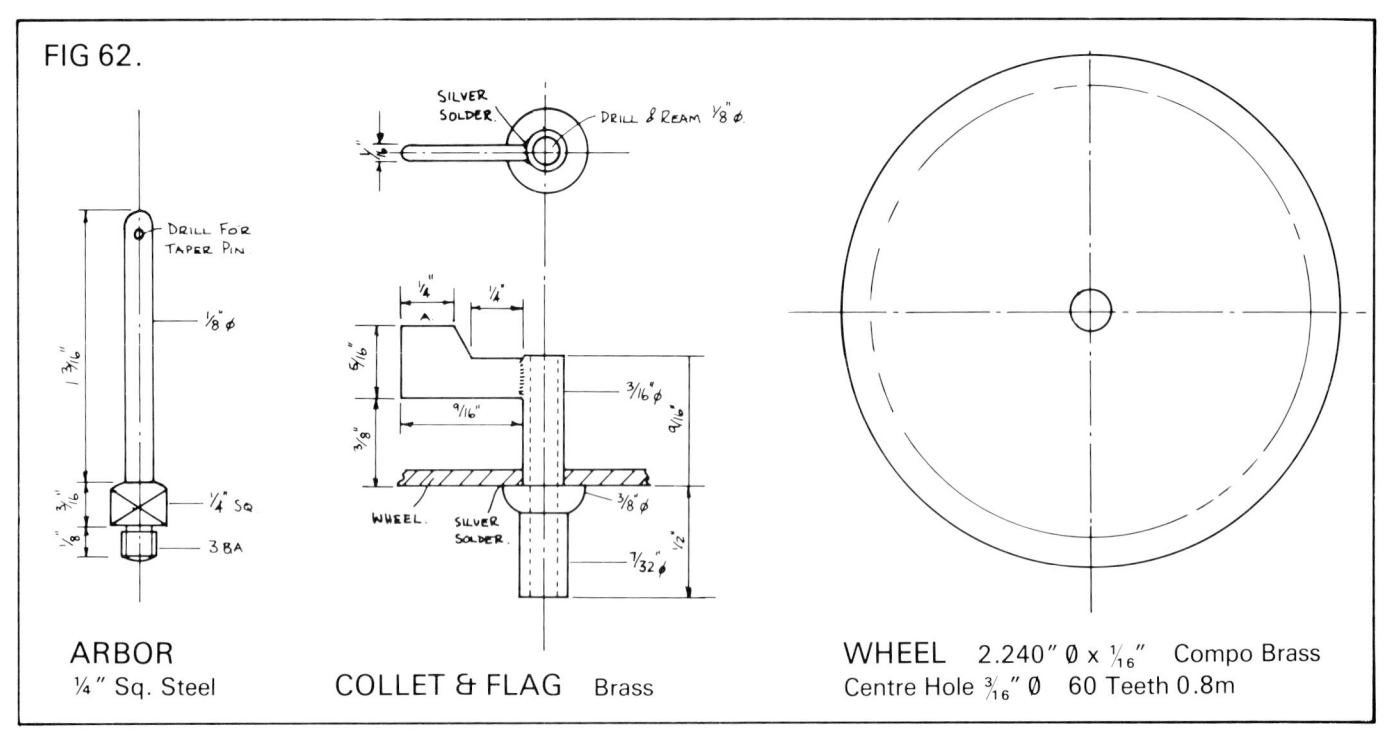

**ARBOR** ¼" Sq. Steel

**COLLET & FLAG** Brass

**WHEEL** 2.240" Ø x ¹⁄₁₆" Compo Brass
Centre Hole ³⁄₁₆" Ø   60 Teeth 0.8m

its seating, and then solder the flag in position.

This wheel is planted on the front plate

*Plate 57. The date ring and rollers.*

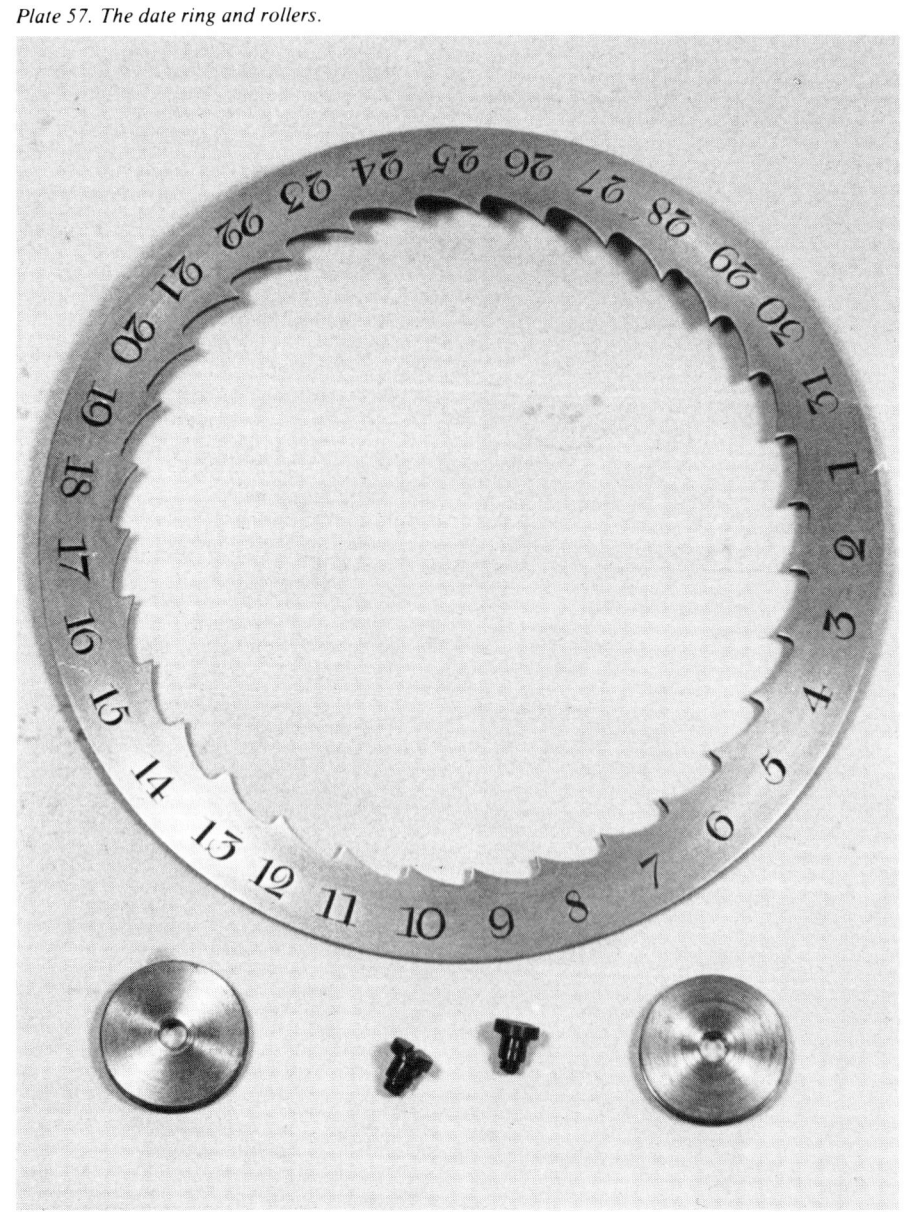

of the movement below and to the right of the hour pipe — looking from the front. On my movement, the centre of the wheel is 1¼" up from the bottom of the plate and 1⅞" in from the right hand edge of the plate. To plant the wheel, set the hour pipe and the date wheel in mesh in the depthing tool, and adjust as usual for the best mesh. Then scribe an arc on the front plate, with the cone runner of the depthing tool located in the centre arbor hole. Then, measure in from the edges of the plate and find the point on the arc which is nearest to the dimensions I have given above. Centre punch this point and drill and tap 3 BA to take the arbor which can now be screwed into place. To fit the wheel, you will have to slide it and the hour pipe over their respective arbors together, as the date wheel is sandwiched between the hour wheel and the snail. Assemble the two wheels to make sure that they mesh well.

**THE DATE RING**

This is the difficult bit! However, there is an easy way out — Goodacre Engraving sell ready made rings with an outside diameter of 6". These rings do not have the numbers engraved, as you have to mount the ring in your clock, draw a rectangle through the date aperture on your dial, revolve the hands 24 hours so that the mechanism advances the ring one step, draw the next rectangle on the ring, and so on 31 times. You then return the ring to the engravers and they will engrave the numbers in the centre of each rectangle — thus guaranteeing that the numbers appear centrally on your own dial. This is the process that has to be done however you decide to do the ring, and for this reason the division of the 31 teeth is not absolutely critical.

The drawing of the ring is reproduced full size, so that you can trace it onto a piece of brass and cut the teeth by hand if you can't get 31 divisions on your dividing head. If you are careful, this will be accurate enough, but I will outline the method that I have used. Cut a square of Compo Brass a little over 6" × 6", find the centre and centre drill. Then drill screw holes in the waste areas at the corners and mount the blank centrally on a plywood disc fixed to the faceplate. You can now mount the faceplate in the lathe and trepan out the centre of the ring as you did with the chapter ring centre. Then the inside diameter can be turned to exactly 4$\frac{19}{32}$" diameter.

Set up a division plate, or dividing attachment, on the back of the lathe mandrel as described previously, or mount the faceplate on a dividing head, so that you can obtain 31 divisions. Now, with a scribing block, or a scriber held with its point at centre height in the tool post, scribe 31 lines outwards from the inner edge of the ring. After this has been done, scribe a line round the ring, $\frac{3}{32}$" in from the inside edge. Remove the faceplate, lay the whole assembly on the bench and centre punch the 31 points where the divisions cross the scribed circle. As the roots of the teeth are about $\frac{1}{16}$" radius, you can drill the centre punched points $\frac{1}{8}$" diameter. At this stage make up a template from thin sheet brass, or tinplate, so that you can scribe the curves between the root circles and the tips of the teeth. The shape of the template can be traced from the drawing. The exact profile is not too critical, as long as the flag clears the backs of the teeth and engages cleanly with each tooth face.

With all the marking out done, return the faceplate to the lathe and trepan the disc out, leaving the outside diameter a little over size. Just before the tool breaks through, stop the lathe and finish off by turning the faceplate by hand. Then, mount the blank in the three jaw chuck with the jaws gripping outwards — holding the ring by its inner edge — and turn the outside diameter to 6". Using a fine file, slightly round the corners of the outside edge.

To finish the teeth, get out your piercing saw and a supply of fine blades, and saw out each tooth to the scribed lines. This is a rather tedious job, but you get there in the end! After sawing, file each tooth to shape with needle files and smooth up with emery sticks. As this tooth shape is quite strong, there is no need to leave a flat at each tooth tip.

Before you can mark the positions of the numbers, you will have to finish the remainder of the parts so that everything can be assembled. The date ring rests on two rollers which are mounted on shouldered screws on the back of the dial. These are shown in the drawing and are simple turning jobs. To ensure that the grooves in the rollers are concentric to the centre holes, mount them on a $\frac{3}{16}$" man-

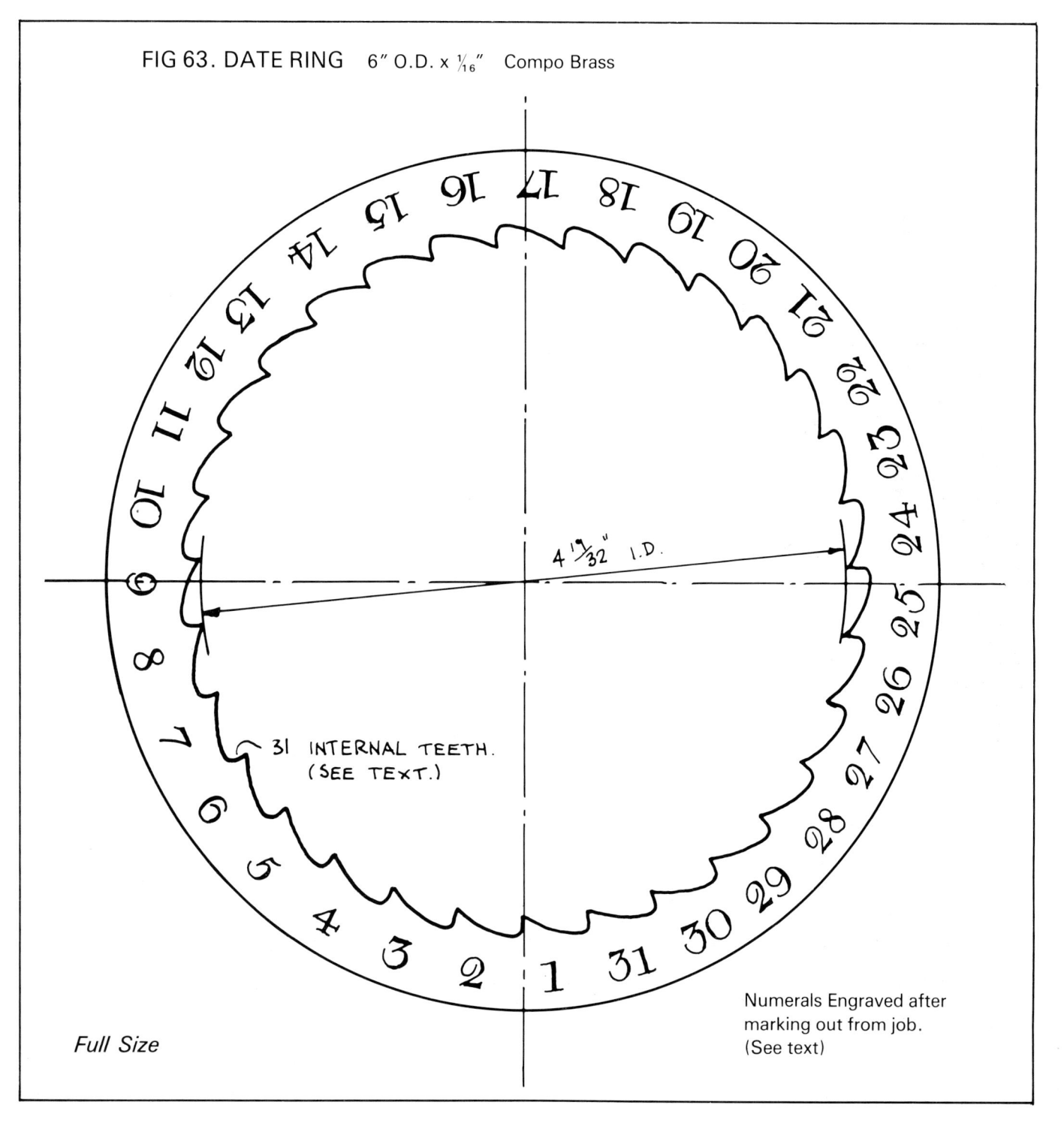

**FIG 63. DATE RING**    6" O.D. x $\frac{1}{16}$"    Compo Brass

4$\frac{19}{32}$" I.D.

31 INTERNAL TEETH. (SEE TEXT.)

*Full Size*

Numerals Engraved after marking out from job. (See text)

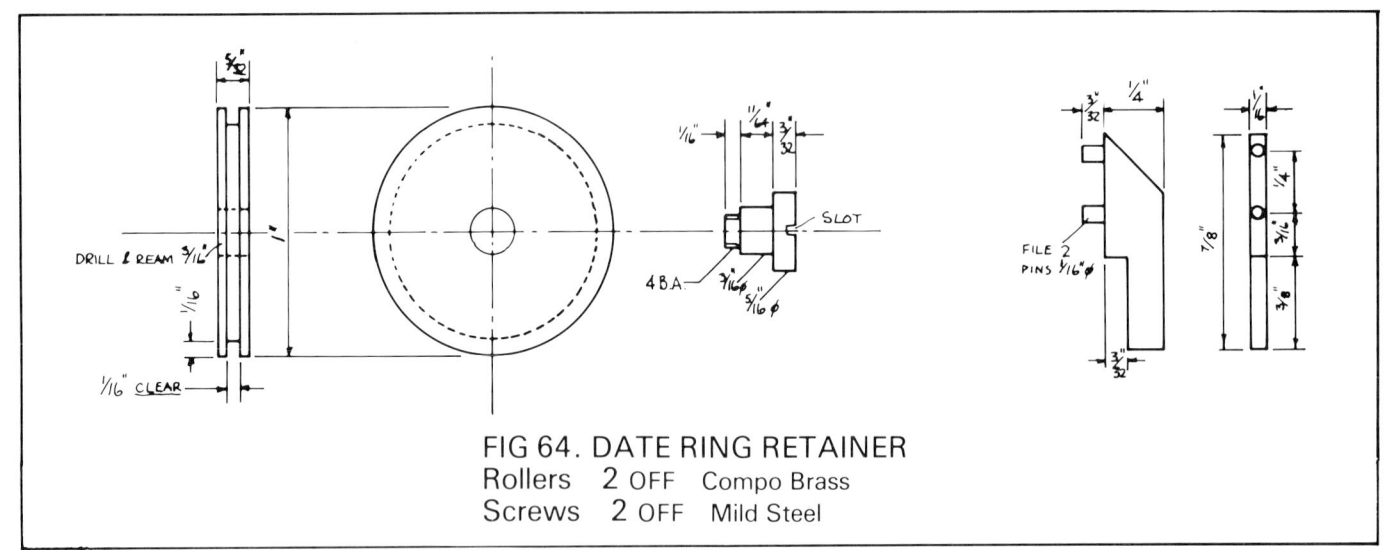

**FIG 64. DATE RING RETAINER**
Rollers  2 OFF  Compo Brass
Screws  2 OFF  Mild Steel

drel held in a collet, exactly as you did when turning the wheel blanks. The grooves should be wide enough to be a sloppy fit on the rim of the date ring so that there is no chance of the ring binding. The rollers should be a free running fit on the screws when they are screwed fully home.

The dial plate is drilled and tapped 4 BA for the roller screws in the positions shown on the drawing. The holes are equally spaced either side of the centre line, $1\frac{11}{16}''$ down from the dial centre. This should put the ends of the screws out of sight behind the chapter ring, so make sure that this is the case before you drill the holes.

The top of the ring is held in place by a retaining 'hook' which is made of $\frac{1}{16}''$ thick brass sheet, and riveted to the dial in the position indicated. With the ring resting on the two rollers, there should be enough clearance for the ring to shake around a little, without it jumping out of the roller grooves. If you wish, you can modify the ring mounting by fitting a third roller at the top instead of the retainer. If you do this, do make sure that the ring is perfectly free to revolve.

The shape and position of the date aperture are shown in the drawing, Fig 65. The top and bottom of the aperture are arcs struck from the centre of the dial. Plug the centre hole with a piece of wood and then use dividers to mark out. Both sides of the hole are radial lines struck again from the centre. After marking out, chain drill round the hole, remove the waste and file to shape. It is usual to chamfer the front faces of the aperture and polish the angled faces to contrast with the matting.

### SETTING UP AND MARKING OUT FOR THE NUMBERS

With all the parts made, assemble the ring on the dial, and the wheels and motion work to the front plate. Then fit the dial to the movement, and try the action of the datework. The front face of the flag should just clear the back of the dial. As we have left the flag a little too long, it will probably bind at the root of a tooth and jam up — file the acting tip of the flag a little, until it just clears the roots of the teeth, and it should then advance the ring one tooth at a time.

Using a sharp pencil, draw through the aperture onto the date ring to mark the shape of the hole onto the ring. Then revolve the motion work 24 hours and repeat the drawing.

Turn the ring back by hand to see how the two rectangles you have drawn relate to each other. The top corners of the rectangles should just touch — or slightly overlap. If they don't, you will have to adjust the length of the flag to obtain the correct throw. When you have got things right, continue all the way round, drawing 31 rectangles on the ring.

The numbers are positioned in the centre of each rectangle. I always send mine to the engravers as they produce by far the best results. The only way I can think of for the amateur to apply the numbers is with a set of $\frac{3}{16}''$ high number punches but, of course, these will not produce such delicate results. The engravers are quite quick — usually about a week — and their

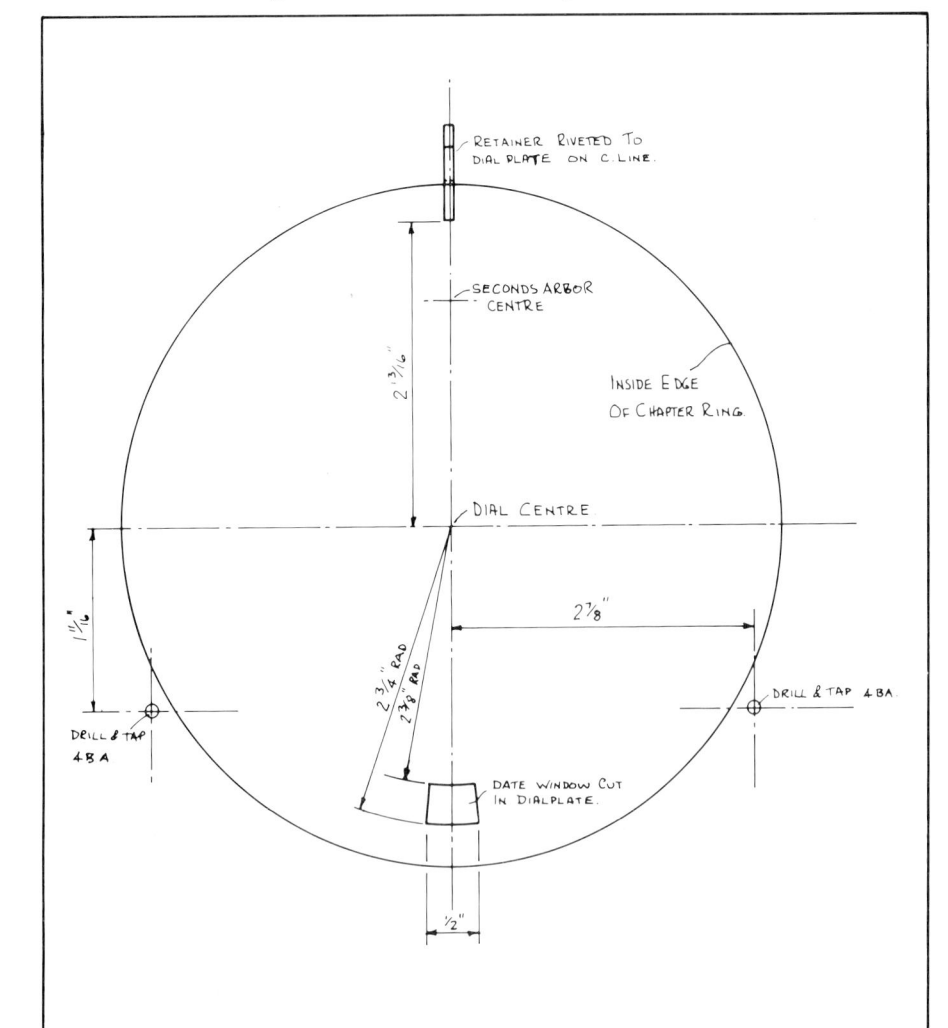

**FIG 65. POSITION OF HOLES & FITTINGS ON DIALPLATE**

charges are very reasonable for this skilled hand done job. Machine engravers may tackle the job, but again, the results will not be so delicate.

When the numbers are done, the ring is given a circular grain, filled with black wax, and silvered in exactly the same way as that described for the chapter ring. It is inevitable that you will get wax into the ring teeth, and it is important to make sure that this is all removed. After silvering, lacquer the ring as usual.

To finally set up the datework, assemble the dial components, oiling the rollers, and check that the ring revolves freely.

Assemble the gears and motion work in such a way that the flag will start to go into mesh with the ring at twelve o'clock. Of course, you now have to remember that this position will always be midnight, so if the clock runs down and stops, you will have to make sure that the date moves at midnight and not midday!

You may find that if your clock is only just managing to keep going without datework, the extra friction encountered when the date is being advanced will be enough to sop it completely. It is important that friction is kept to the absolute minimum, and so if your clock stops on the change, find the tight spot and free it off.

I hope that is all clear enough, and that many of you now have a movement that is nearing completion.

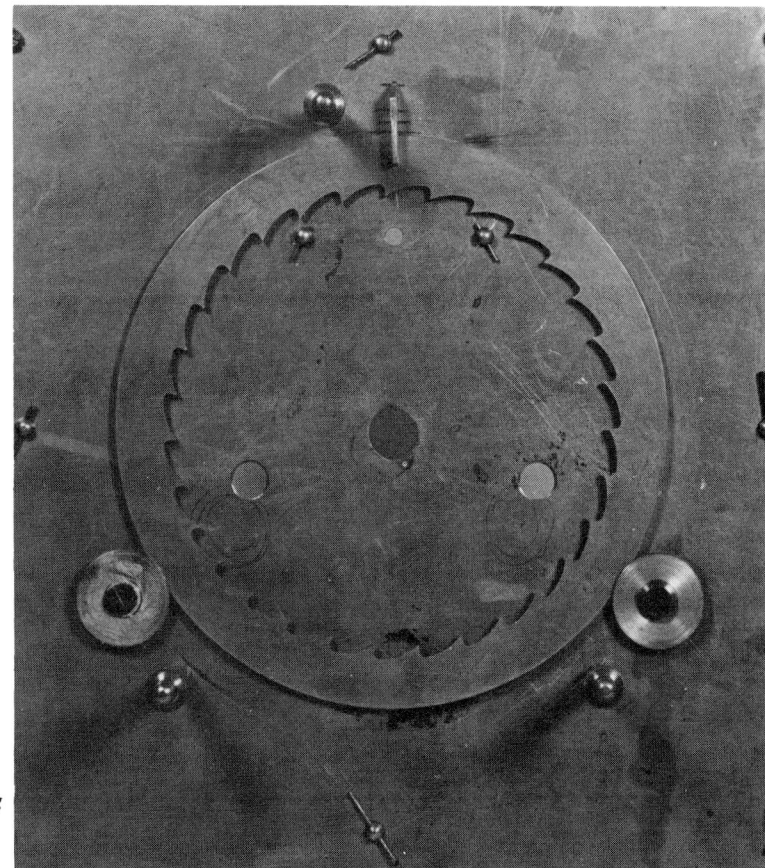

*Plate 58. The date ring fitted to the back of the dial plate.*

77

# Chapter 15

**FINISHING AND CLEANING**

Now that all the parts have been made, and the movement has been test run, you can concentrate on the finish, prior to final assembly. The main point here is that all the parts should be absolutely clean and dust free. It is surprising how little dust it takes to stop a clock — especially if the dust is in the wrong place! On the other hand, I am often amazed at the way some antique clocks run, even though they are covered with vast amounts of filth oily dust which has collected over the years. Of course, the other problem is that dust and dirt causes a lot of wear. I think I mentioned earlier that it is usually the steel parts that wear, not the brass. This is because the abrasive dust beds into the softer brass which then acts as a lap, and gradually laps away the harder steel parts. Worn pinion leaves are a major cause of faults on old clocks. I have had clocks in my workshop for restoration with their pinion leaves cut right through, but with little signs of wear on the wheel teeth. If you don't want this to happen to your masterpiece, make sure you really do thoroughly clean everything.

You must also decide wether you want a polished finish as I do, or a good scratch free semi-matt grained finish. In a longcase clock, the movement is not usually visible, so a polished finish is not so important as it is, for instance, on a skeleton clock under a glass dome where all the movement is seen. If you intend to enter your clock for competition at exhibitions, which I hope some of you will do, I think the judges usually like to see a good polish on all parts — and this certainly looks by far the best. Polishing also tends to delay tarnishing of the brass.

After testing the movement for at least a full week, strip it down completely. Of course, if there are any problems during the test run, sort them out now, before you add the finishing touches. Put all the pieces in tins or boxes so that nothing gets lost, taking care not to damage wheel teeth etc. I get my wife to save the plastic tubs with press on lids that are used for ice cream at freezer food shops. These are ideal as they are airtight and prevent condensation causing rust on stored parts and also, being made of polythene, they do not cause damage to delicate teeth etc.

If you have followed all the instructions so far, you should have a collection of bits and pieces which are all basically finished, with no scratches or machining marks. If you have got scratches on any parts, now is the time to get rid of them. Do this with very fine wet and dry paper, either wrapped round a block of wood or glued to sticks. Again, always take great care and avoid rounding off sharp corners and edges. If you have any bad scratches in the centre of the plates, you will find it easier to remove these by careful use of a small scraper. Finish the plates with wet and dry paper rubbing in straight lines up and down their length to produce a very fine straight grain. All other flat parts should be given the same treatment to obtain the best possible finish before you start polishing. Make sure that all the wheel crossings are nicely finished (it's surprising how easy it is to miss one out), and spin the arbors in the lathe to polish them up, again using fine emery sticks.

When you are sure that all the blemishes have been removed, you can begin polishing. There has been a good deal of argument in the horological world about polishing methods for clock movements. This argument and discussion has arisen because of damage caused by mechanical polishing methods, ie, the use of buffing machines. In my opinion, if a buff is used carefully, using the correct materials, no harm will be done. If parts are buffed up on a hard mop with lashings of coarse polishing medium such as brown 'Lustre' and heavy pressure, then it is manifestly obvious that corners will be rounded off and wheel teeth will be damaged. Buffing is not intended to remove bad scratch marks. It is only intended to impart the final shine to an already bright surface.

The only mops which I use are the very soft 'Swansdown' mops, and these are used with the very finest grade of jewellers rouge, which is obtainable in stick form ready to apply to the mop. Even this fine rouge is only used very sparingly. I also find rotary bristle and fine brass wire brushes are very useful and do not cause damage. The lucky ones amongst you may also have access to a jewellers pendant motor and flexible shaft. These have collet handpieces to take standard shanked tools. There are a huge number of cutting and polishing tips available for these tools, and the polishing tips available are most useful for polishing in awkward places and inside wheel crossings, etc. The small twelve volt mini-drill type tools are also very useful for this type of job, and these are somewhat cheaper than the professional jewellers motors.

The traditional polishing method was to use some form of polishing medium on a bristle brush. A one time common polishing medium was rottenstone, a mild abrasive, which was mixed with oil and then brushed on vigorously with the brush until all trace of the polish had disappeared. I don't know exactly what rottenstone is, or even if it is still available. However, I do know that one of the modern metal polishes such as 'Brasso' or 'Solvol Autosol' works well with the brushing method. The polishes with a creamy paste consistency work best as it is easier to charge the brush without the polish running and splashing all over the place. All the parts should be brushed vigorously until they start to shine. Pinion leaves are polished with a pointed piece of wood charged with fine abrasive. If you rub hard, the wood will take up the form of the pinion leaves and will not alter the shape of the pinion leaves.

Oil sinks and pivot holes are cleaned with pegwood which is available from clock tool and material dealers. The pegwood is simply sharpened with a sharp knife, and is then pushed into the hole and twisted with the fingers. When the wood is dirty, re-sharpen it and repeat until the wood comes out clean. Make sure that all the holes for screws and register pins are quite clean otherwise there is a danger of polishing medium being pushed out on assembly.

When all the parts have been polished, using whichever method you have chosen, they should be washed. Benzene or petrol was originally used for this, indeed, many people still use them. Petrol has the advantage that it evaporates from the job, leaving it dry, but it goes without saying that it is a very dangerous method requiring very great care. Petrol is also harmful to the skin as it removes natural oils leaving the skin very dry. There are several proprietary cleaning fluids on the market nowadays, most of which are very good. The most effective cleaning method is to use one of these fluids in an ultrasonic cleaning tank. The parts are immersed in the tank which has powerful transducers glued to its underside. These emit a powerful ultrasonic vibration which really does get at all the dirt, vibrating it out of all the nooks and crannies, and at the same time aiding the cleaning properties of the fluid. Unfortunately, these tanks are expensive and are only worth buying if you are cleaning a lot of clocks.

However, some of the cleaning fluids work well without a tank. Those that I can recommend are 'Horolene', — an ammonia based concentrate which is mixed with seven parts water — 'Dipbright' used neat, and 'Walkers Well Clean' and 'Well Dry', which are spirit based fluids containing no water. 'Well Dry' evaporates off the job leaving it perfectly dry, and is used as a dip after the cleaning process.

If you wish, you can make up your own brass cleaning fluid to the old clockmakers formula. This involves dissolving about 2 oz. soft soap in a couple of pints of water which is heated to almost boiling point. It must not be allowed to boil. (Soft soap is a green jelly available from the larger chemists). Let the liquid cool and then add about three large tablespoons full of 80 Ammonia — this is the very powerful full strength stuff, and gives off the most awful fumes, so take care and mix it up outside.

To use these fluids, put the parts to soak for about a quarter of an hour, making sure that they are totally immersed in the liquid. If any brass sticks up above the surface of the liquid, a nasty brown tidemark will appear, and this is very difficult to get rid of. After letting the parts soak, put on some rubber gloves and give each piece a good brushing using a soft bristle brush. These brushes are available quite cheaply from material suppliers. When the parts are thoroughly clean, rinse them preferably in very hot clean water. If you use a spirit based fluid such as 'Well Clean', you must use the recommended rinse and not water. After washing in water, drop the parts into a container full of Boxwood dust. This is a very fine sawdust available from some horological suppliers, and is the traditional material for drying washed parts. Leave the parts in the dust for about half an hour when they will be perfectly dry. Pull them out one by one and tap the excess dust back into the box.

Finally, to get a really bright shine and remove all traces of dust, give all the parts a good hard brushing with a soft clean brush, rubbed on a block of chalk which will act as a superfine polish. Brush each part until it gleams and carry on brushing to get rid of all trace of chalk. Do each part in turn, re-charging the brush with chalk as you go. At this final stage. it is important that you do not touch the brasswork with bare fingers, so wear rubber gloves. As I have mentioned before, I use very thin surgical gloves which are available in boxes of 25 pairs and are cheap, so you can afford to throw them away when they get dirty. They are packed in boxes like tissues under the brand name of 'Kimguard'. You will have to check locally for suppliers.

As each piece is finished, place it carefully in a box with a lid to keep the dust off. Peg out all the holes again, just to make sure all the chalk is removed, and prepare the final set of screws and pins, ready for final assembly. Polish all the screw heads in the lathe, using fine emery sticks. If you wish, the appearance of the movement can be enhanced by blueing all the screws. This will also help to prevent tarnish or even rust later.

**FINAL ASSEMBLY**

There is no need for me to describe the positioning of components as they will have been in and out of the plates so many times by now that they should be able to assemble themselves! I will merely give a few tips on the things to watch out for, etc.

Assemble all the wheels, arbors etc., between the plates, taking care that the pins on the strike train are in their correct relative positions. At this stage, do not oil any of the pivots. Assemble everything dry. Make sure you use suitable sized screw drivers to avoid damaging the screw heads, or even worse, slipping off a screw and scratching other components.

When you fit taper pins, it is neater to cut them down to a reasonable length with no excess bits poking out. Cut the pins with small end cutters and then take the trouble to round off the cut ends with a fine file.

When you have assembled the two trains, double check to make sure that they are free running, with no tendancy to stop suddenly. If all is well, apply a tiny drop of oil in the oil sink of each pivot. I use a small hypodermic syringe for oiling, but you can just as easily use a needle dipped in the oil to apply a couple of drops to each pivot. Use only the correct clock oil which is available from the suppliers already listed, and do not be tempted to overdo the oiling.

Next, assemble the motion and lever work on the front plate, applying a little oil to the stub arbors, the tips of the friction spring and the inside of the hour pipe. Make sure that you mesh the two minute wheels so that the minute hand does point to twelve as the strike train is released. Check the action of the rack spring and make sure that the rack tail engages at about the middle of each snail step when the rack drops. Also check that the warn pin engages correctly with its stop when the rack drops. Determine the correct position for the gathering pallet and fit it to its arbor, using a drop of Loctite retaining compound. When everything is together, test the strike work a few times to make sure it works perfectly. It is best to find any faults at this stage before the dial is fitted.

With the front plate assembled, fit the pallets and the back cock, not forgetting to put a tiny drop of oil on the pallet faces and the pivots. Test the escapement to make sure it works well. If you haven't already done so, fit the lines to the barrels making sure that there are no loose ends sticking out of the barrel between the great wheel crossings. Finally, fit the dial and hands, and there it is at last!

To run finished clocks out of their case, I have made up a 'Perspex' box with three sides and a top which fits over the movement and bell, butting up to the back of the dial. This allows you to see exactly what is going on, but it keeps all of the atmospheric dust out of the works. If you don't wish to make such a box, find a cardboard box of a suitable size and cut it to suit.

# Chapter 16

**CASEMAKING**

Case making is always a problem for amateur clockmakers wishing to make a complete longcase clock themselves. If, like me, you are not particularly keen on woodwork, the sight of a complicated clock case full of mouldings is often enough to make one think twice before starting such a project. Those amongst you with cabinet making or joinery experience will not be so worried. If you really can't cope with the case, there are various complete kits on the market which make a very good starting point.

When I first started this series, I only intended to show a simple oak case to take a square dial, but many people who have seen my clock in a mahogany break arch case, asked if I would also give details of the more complicated version as well. I should stress here that those who asked for this have had some considerable experience of cabinet making, or intended to have a case made professionally. The latter is of course an expensive exercise! Thus, I have decided to provide details and drawings of the easier oak case with instructions on making it, and also to provide outline drawings of the break arch case with all the major dimensions to provide sufficient information for the experienced woodworker to complete the case.

Again, I must stress that I am no cabinet maker, so the methods that I shall describe will be a simple means to an end, and may not be the theoretically correct methods! Please bear with me and adapt or improve as you think fit, depending on the equipment you have to hand.

As a final word of encouragement to the budding casemaker, the vast majority of antique clocks are 'country made' as opposed to London made. If you examine the internal construction of a country case, you will be amazed at the crude workmanship and finish. The cases were put together with nails and all sorts of glue blocks using the simplest methods — but the external look of the finished item is quite acceptable.

The most difficult parts of the case to make are the mouldings, and these will be available as complete kits of all the mouldings needed for each case. Some of the clock suppliers also stock a good selection of useful mouldings, some of which are peculiar to clockmaking — so a look at a few catalogues will be useful. These same suppliers also stock cast brass capitals, and special hood and case door hinges as well as locks and escutcheon plates.

**SELECTION OF TIMBER**

The first decision to be made is the choice of materials. For both cases, the best is undoubtedly solid timber — oak for the square dial and mahogany for the break arch. Solid timber is also the most expensive material, and it may also be difficult to find really good quality, properly seasoned wood. Of course, on the mahogany case, the beautiful 'flame' on the door and the base panel is mahogany curl veneer, which is fairly easy to obtain. The veneer in this case is applied to either an oak or mahogany base.

The selection of timber for the trunk door is particularly important as this is a thin, unbraced panel which is liable to warp. On an oak case, the best material is good quarter sawn oak — if you can find a good piece. A cabinet maker friend of mine is always on the lookout for old furniture such as bedheads etc, which are often good quality, dry timber. This is a very good source of supply of good quality well seasoned timber, and possibly the cheapest — so keep your eyes open, or visit a few house clearance sales. I know several people who have made complete cases from old furniture, and very nice cases they are too!

Another method of case-making is to build the carcase from blockboard, which is stable. This is then veneered and solid mouldings are applied. This is a good method, but it involves much more work, and can be tricky if you are not well practised in veneering. Remember that both sides of the panels must be veneered to prevent warping.

The internal frames and other unseen parts can be made of softwood which is easily obtainable, and keeps the cost down. This is also used for glueblocks and possibly for the backboard.

**THE OAK CASE**

Hopefully, the drawings are fairly self explanatory. As you can see, the case consists of three main sections — the base, the trunk and the hood. The base and trunk are built as two separate sections and then joined together. The hood slides onto the trunk from the front, thus giving access to the movement. The hood also has a glazed door so that the clock can be wound and adjusted without removing it, and there is a door to the trunk for access to the weights and pendulum. The movement and dial sit on a seatboard which rests on the sides of the trunk which extend up past the top moulding into the hood.

The basic material for most of the case is oak boards, ⅝" thick. Buy these planed to thickness if possible. If you find some old oak, then you can get away with ½" thickness if you adjust the dimensions to suit where necessary. If you obtain thicker material, a local joinery firm will often help out and machine the timber to the required thickness. Before you ask them, remember to check very thoroughly for any pins or nails, as the machinist will not take kindly to having planer blades hitting unseen metal.!

Before you begin cutting the wood, I strongly advise all constructors to obtain a large piece of hardboard and some white 'lining' wallpaper. Paste the paper on to the board, and draw yourself a FULL SIZE front and side elevation of the case, including any relevant cross sections. This is common practice, and certainly makes sense as you can then take accurate measurements from the drawing. This also gives you the opportunity to superimpose front and side elevations of the dial and movement, measured from your own, complete with weights and pendulum. From this, you will soon spot any areas where a little more clearance may be needed, as well as determining the correct position for the seatboard to suit your own movement. The drawing will also serve to familiarise the builder with the construction methods used, so that he can make a start with a clear view of what lies ahead.

The best place to start is the front trunk frame, which is shown in Fig 67. This is a fairly straightforward job, but it must be true and FLAT. Before you start assembling parts of the case, you are well advised to make up a large FLAT worktop. I find that a hardboard panelled door serves very well, and these are not too expensive.

The front frame should preferably be made up using mortice and tenon joints, or alternatively it can be dowelled using one of the dowelling jigs available from do-it-yourself shops. Glue the frame up with one of the modern resin adhesives. You will see from the cross section that the front frame is grooved to take tongues on the edges of the side panels, and that the back edges of the sides are rebated to take the backboard. This is the strongest method of construction, but if you really cannot cope with this, then these joints can be butt jointed and reinforced with pins on assembly. If you do this, it is most important that the straight edges are planed really straight to make the strongest possible joints. The joints are further strengthened with triangular glue-blocks down the inside corners. Remember when cutting the sides that these project above the top of the front frame to take the seatboard. Either cut them to exact length, checking this from the full size drawing, or alternatively leave them a little over length and trim them to fit the dial to the mask later. The bottom ends of the sides and front frame should line up exactly. After glueing the sides to the front, tack a couple of strips of wood across the back of the trunk to strengthen the assembly until the backboard is finally fitted.

The basis of the base unit is two softwood frames, and these are shown in Figs 68 and 69. The easiest way to make these is to use half lap joints at the corners, glued and screwed. The inside width of these frames must be exactly the same as the outside width of the trunk, as the trunk will finally fit inside the top base frame. The external dimensions of the

*Plate 59. The oak case.*

*Plate 60. The finished break arched version of the clock in a mahogany case.*

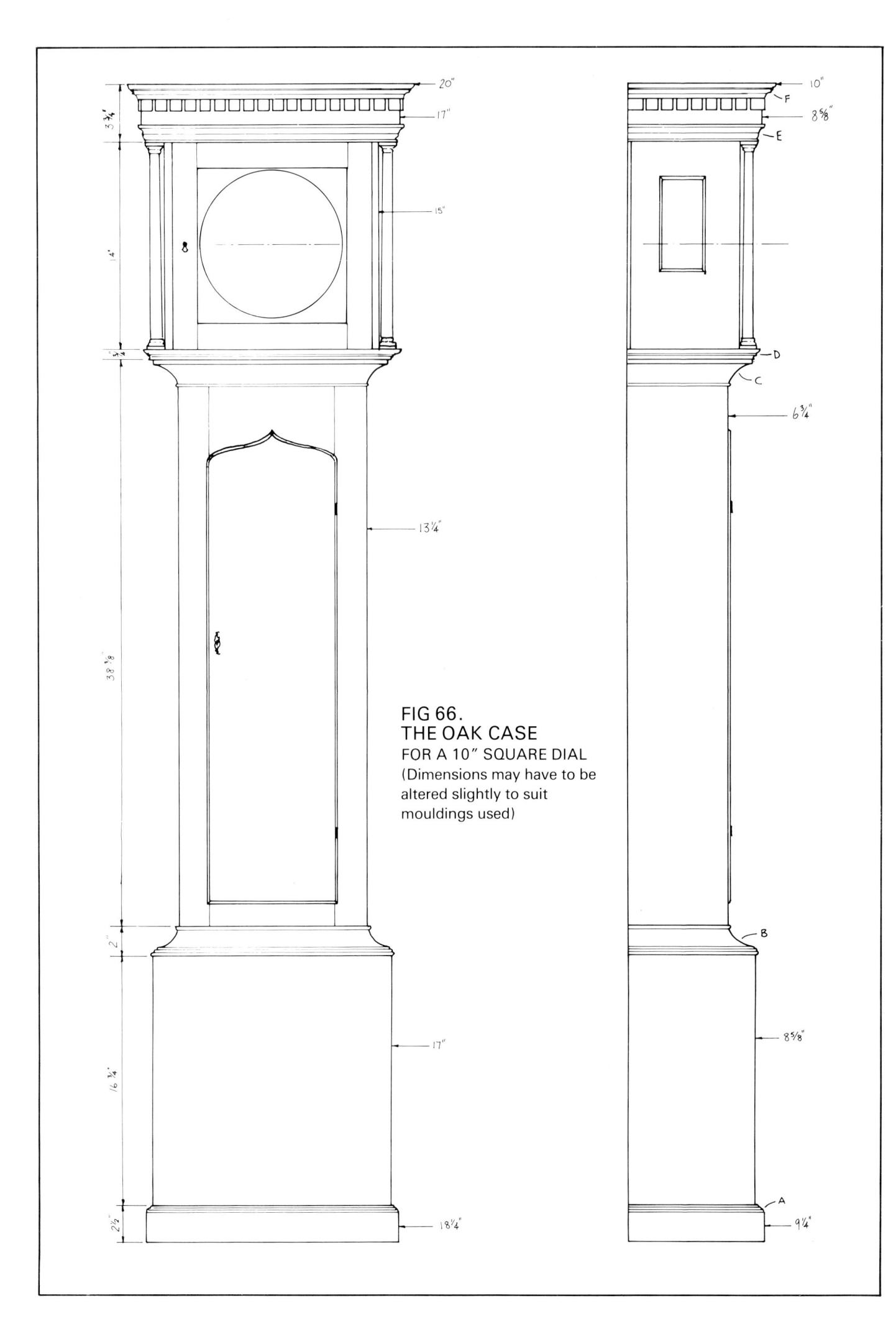

**FIG 66.**
**THE OAK CASE**
FOR A 10" SQUARE DIAL
(Dimensions may have to be altered slightly to suit mouldings used)

frames will depend on the mouldings obtained to fit round the outside of the bottom of the trunk, and so it is best to obtain the mouldings first and then make the frames to suit.

With the frames made, the three panels which make up the sides and front of the base can be cut to size. The front panel will probably need to be jointed from two boards with a good butt joint. The side panels are joined to the front in the same way as the trunk, and again the back edges of the sides are rebated to take the backboard. These panels are glued to the two frames as shown in the drawing and screwed to the frames from the inside. This makes up a second rigid box. The plinth moulding can be fitted later.

When joining the trunk to the base, it is most important that the two line up perfectly in all planes, otherwise the case will appear distorted. (This is where it will be found to be most beneficial to have a full size drawing and a flat bench). Fit the trunk into the top frame of the base so that the bottom edges of the trunk are flush with the bottom edge of the frame all round. Check to ensure that everything is in line and draw a pencil line round the trunk where it enters the frame to aid alignment later. Pull the trunk out of the base and drill about half a dozen clearance holes for woodscrews through the bottom of the trunk — these holes should be about half way between the bottom edge of the trunk and the pencil line which you have just drawn round it, and equally spaced all round the trunk base. After drilling the holes, re-assemble the trunk and base dry, and fit all of the screws. You can then stand the trunk up to check that it all lines up nicely from all angles, and of course make any corrections before applying the glue. When you are perfectly satisfied, separate the two, apply glue to the mating faces, lay the job flat on the bench and fit as many of the screws as you can reach before turning the case over to fit the remainder. Check for alignment once again before leaving the assembly to dry. At this stage, the job should begin to look something like a longcase!

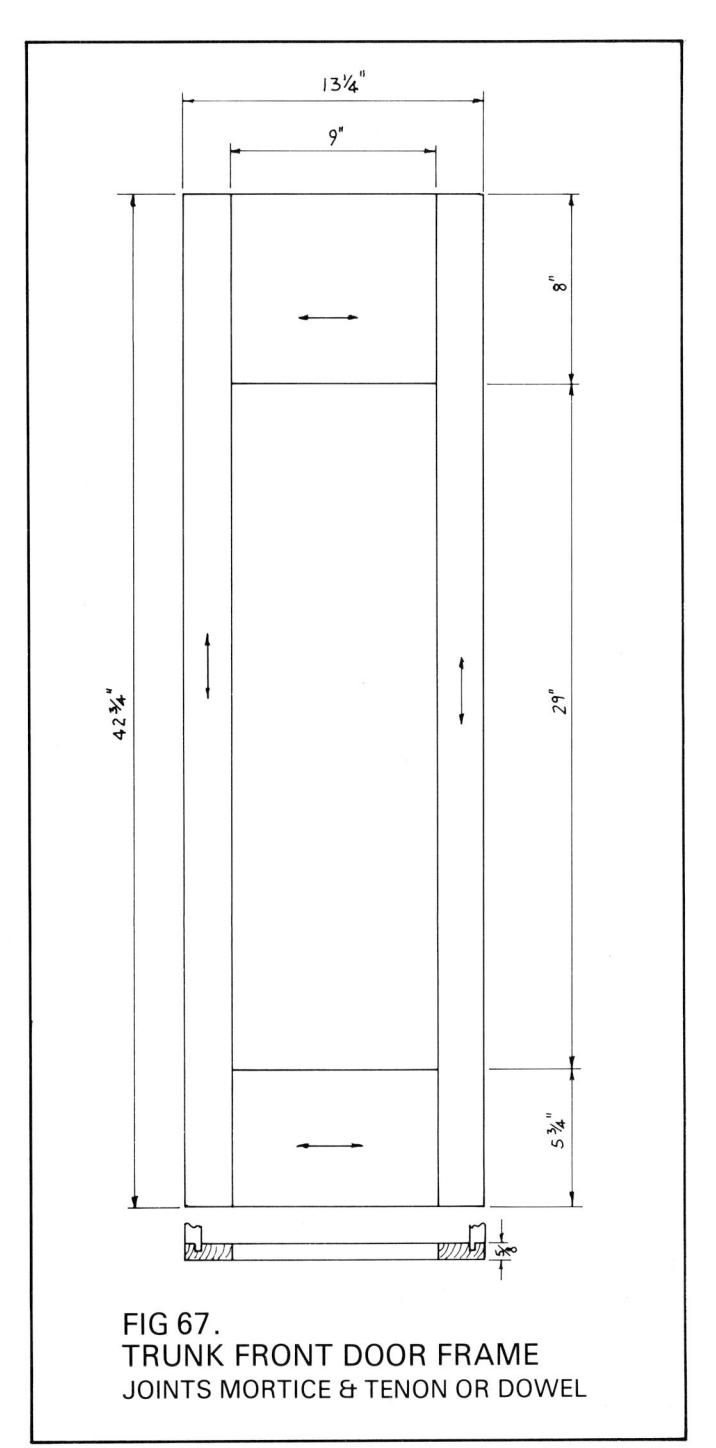

**FIG 67.**
**TRUNK FRONT DOOR FRAME**
JOINTS MORTICE & TENON OR DOWEL

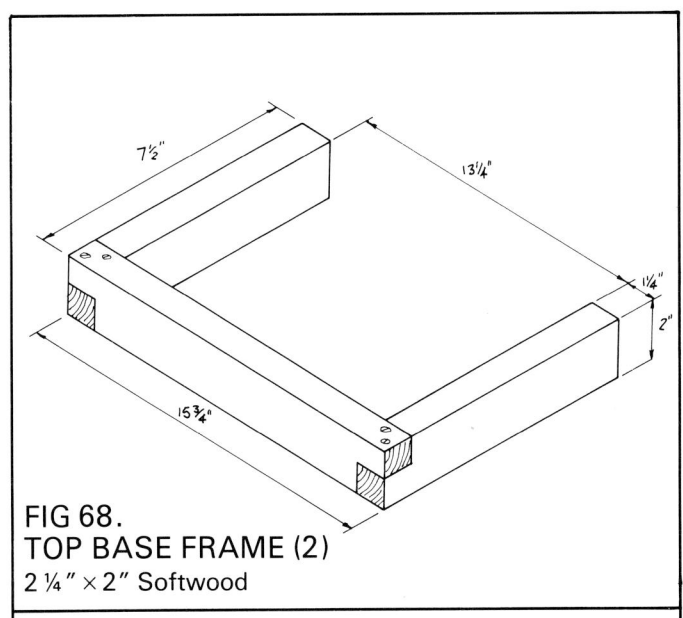

**FIG 68.**
**TOP BASE FRAME (2)**
2 ¼" × 2" Softwood

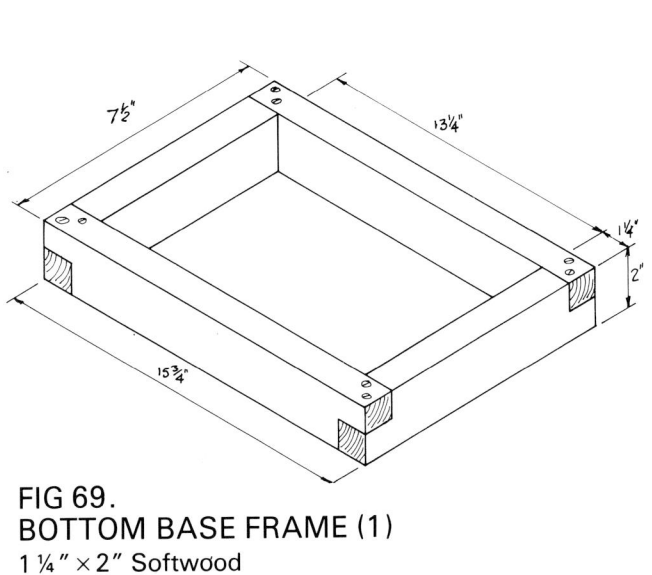

**FIG 69.**
**BOTTOM BASE FRAME (1)**
1 ¼" × 2" Softwood

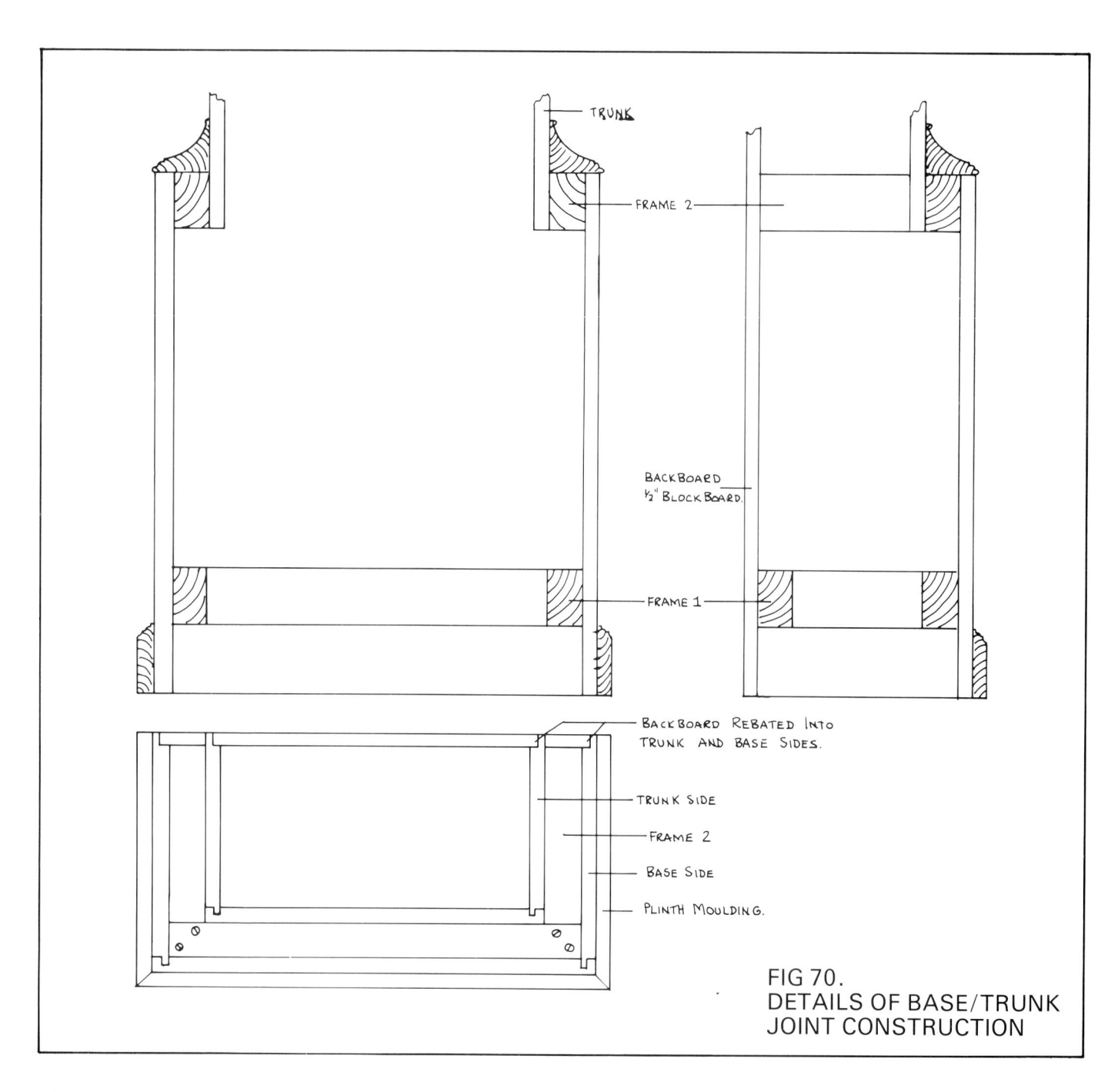

FIG 70.
DETAILS OF BASE/TRUNK JOINT CONSTRUCTION

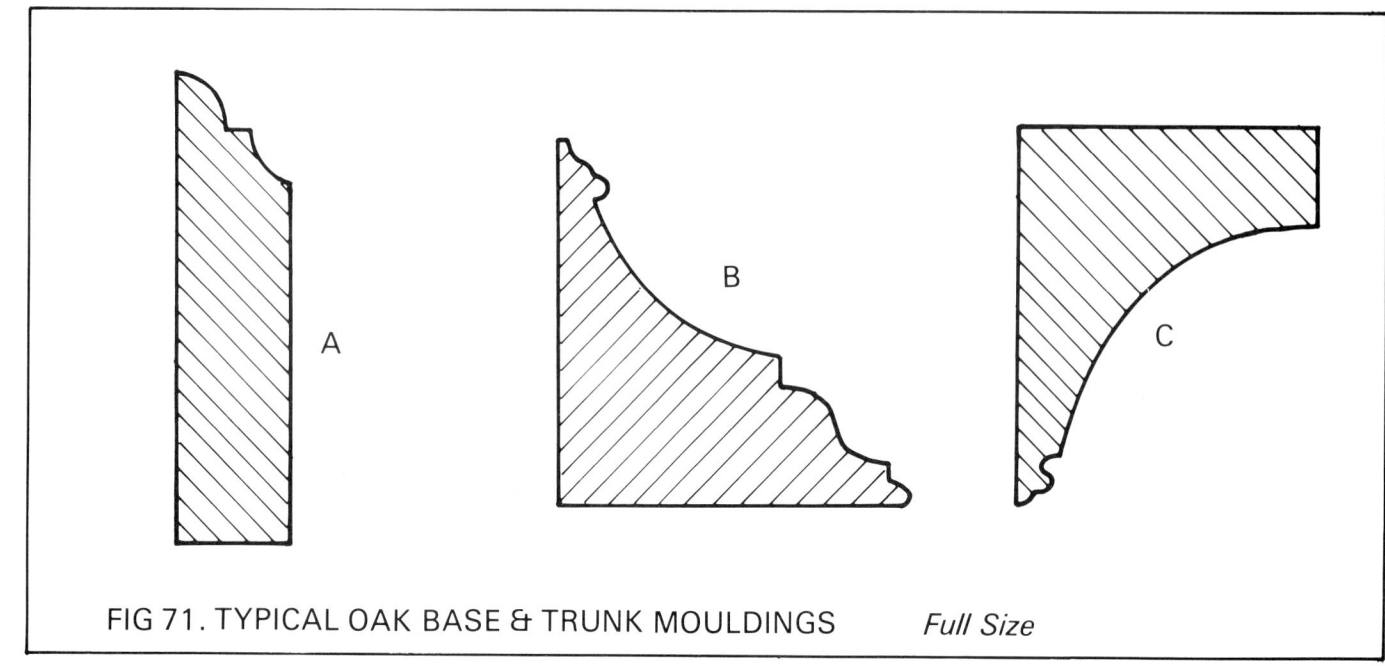

FIG 71. TYPICAL OAK BASE & TRUNK MOULDINGS  *Full Size*

**FIG 72.**
**TRUNK TOP & SEATBOARD DETAIL**

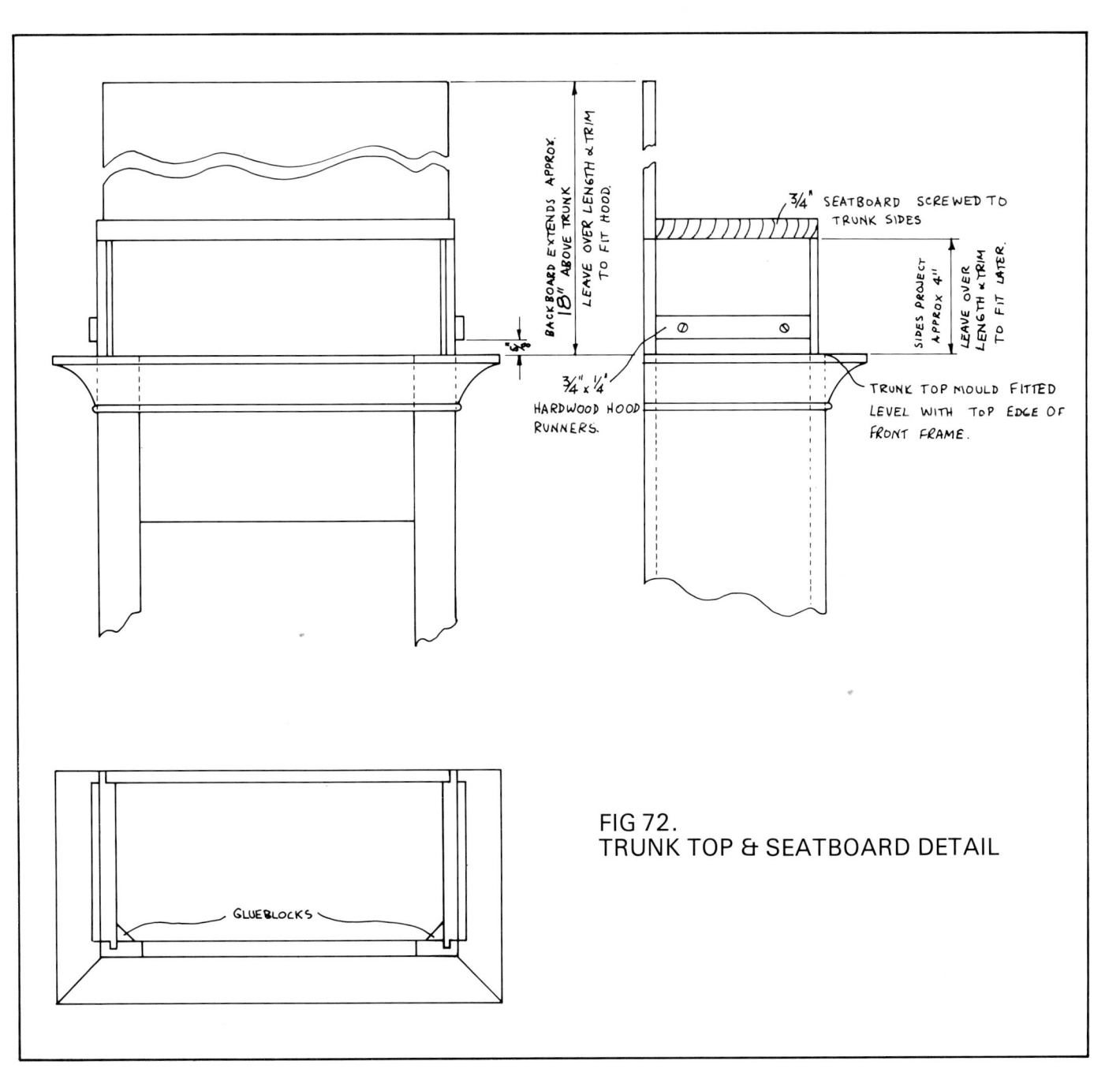

Typical case mouldings are shown full size in cross section. They can be planed from the solid, or alternatively, the more complicated ones can be built up — the end result will be the same. Commercial mouldings are made on a spindle moulder which few amateurs will possess. Some of you, however, may own a small circular saw table or a router — both of which may be used to form the mouldings. The age old method of course is to use wooden moulding planes with shaped blades and bodies, finishing the shape off with a scratch stock. If you are careful (and patient) you can shape all the smaller mouldings with a scratch stock alone. The blades for scratch stocks are simply made using pieces of old HSS hacksaw blades ground to the required shape, or thin gauge plate which can be filed to shape and hardened. The blade is then fitted to a wooden stock which acts both as a depth stop and fence. To use a scratch stock, shape the blade as required, sketch the section required on the end of a suitable piece of timber, plane off as much of the waste as possible, and scrape the final shape using the stock in the same way you would use a cabinet scraper. The mouldings are then finished using glasspaper on suitably shaped pieces of scrap wood.

If you decide to buy your mouldings and you cannot obtain them exactly as shown, do not worry as the ones shown are only examples of the many different styles used. All you need to do is to make sure that the dimensions of the building frames are altered to suit your mouldings.

The corner joints of all the mouldings are mitred. Take great care to cut exactly to the correct lengths and make, or buy, a good accurate mitre box. Gaps in the joints look awful, and are difficult to fill effectively. The larger mouldings are glued in position and screwed from inside the case, especially the trunk top moulding which will eventually carry the weight of the hood, and so this must be firmly attached. The smaller mouldings can be glued and pinned in place and the pin heads punched in and hidden with filler of a suitable colour.

The moulding round the base of the case which forms the plinth can be left flat along the bottom, or alternatively it can be sawn to form decorative feet. A point to bear in mind is that if the clock is to finally stand on a carpet, as large an area of 'foot' as possible will be an advantage, as this will prevent the clock from sinking into the pile of the carpet (possibly unevenly) — this is a common cause of clocks stopping when they are first set up. If the clock is to stand on a firm surface such as floorboards, there is no danger of settlement and so small feet will not be a disadvantage.

With the base and trunk assembled, you can now fit the backboard temporarily. On antique clocks, the backboards are almost always rough, warped, worm eaten planks of either pine, elm or sometimes oak. Very often, the backboard is a source of trouble particularly it if has split (very common) or warped and pulled the rest of the case with it. Thus, the back is an important structural part of the case. If you can get hold of some well seasoned pine boards, these should be jointed together, planed smooth and cut to fit the rebates in

FIG 73. HOOD DETAILS

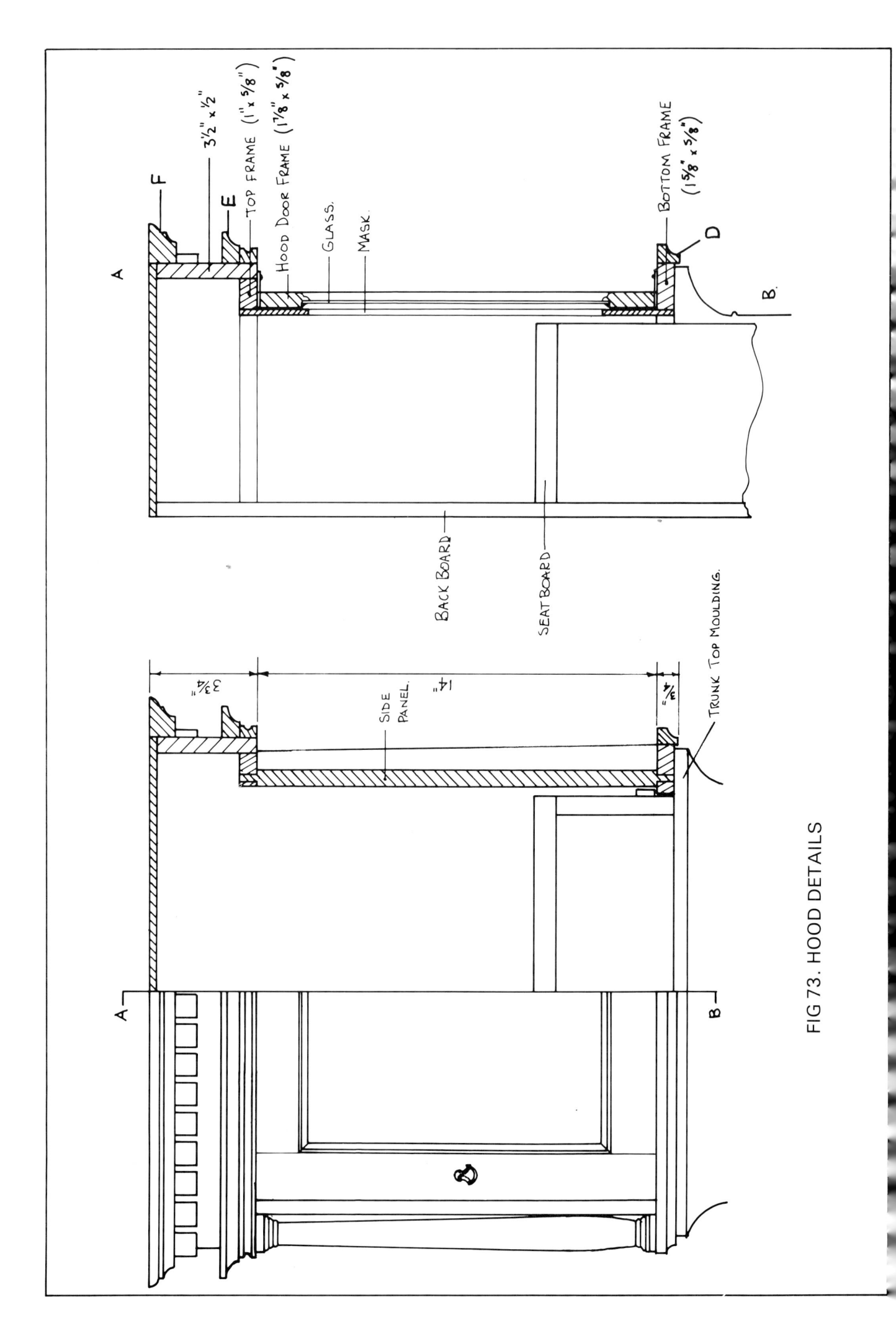

the sides and base. When you do this, check to ensure that the sides of the case have not sprung either in or out. The backboard should project about 18" above the top of the trunk moulding to form the hood back. The width of the backboard across the hood increases to about 15". As we want the best possible fit between the hood and the back to keep the dust out, leave the top end of the backboard oversize so that it can be marked out accurately when the hood is in position. On completion, the inside of the backboard can be stained to match the inside of the case if you wish.

Probably a better choice for the backboard is a piece of ½" thick oak faced ply, or ½" blockboard if you can get it. The reason for this is because plywood or blockboard can easily be bought in one piece, it is flat, and most important, it is stable. Thus, it will be much easier to make and a more reliable alternative and it will not show. I advise fitting the backboard temporarily with pins so that it can be removed later for trimming to fit the hood. When this has been done, glue and pin the back into place very firmly. Take care over this and the case will hold together for many years without distortion.

The final construction work on the trunk is to cut a dozen or so triangular glue blocks and rub these into place, spaced at intervals down the inner front corners of the front frame and sides.

## THE HOOD

This is the trickiest part to make, and it must be really well made because it is open backed, and thus not as structurally sound as the rest of the case. The hood slides backwards onto the top moulding of the trunk, and is located by the rebated moulding around the hood base. I have shown this moulding as a built-up frame with a moulding applied round the edge to form the rebate for ease of construction. These mouldings are available machined from the solid if you prefer to buy them.

The main frame of the hood consists of the two sides and the mask which will surround the dial like a picture frame. The mask fits into grooves cut in the inner faces of the side pieces and it projects ⅝" above and below the sides. This stage of the assembly is shown in the sketch, Fig 75. The mask is rather thin and so it is best

*Plate 61. The hood.*

made from ¼" thick oak veneered ply. The hole in the mask should be of such a size that it will overlap the edges of the dial plate by about ⅛" all round, so check this from your dial.

This first stage is rather critical as the hood must be a nice sliding fit on the trunk so that it can be positioned easily, and the mask must fit up closely to the dial plate with the movement in position with the pendulum bob clear of the backboard. On my clock, the front face of the dial plate is some 6⅜" away from the front face of the backboard. Your full size drawings should tell you if this is satisfactory for your clock, but I advise you to make a ¾" thick seatboard with holes cut in it in exactly the same places as those on your clock horse. Fit this seatboard in its approximate position on the top of the trunk sides and try your movement in place with the pendulum attached. Better safe than sorry!

As you will see from the cross section of the hood, the bottom frame is a close fit round the trunk sides and the hood sides are about ¾" wider inside. This allows for two runners to be screwed to the trunk sides which will prevent the hood from tipping forwards. Thus, the inner faces of the hood sides should be 14" apart. The overall size of the mask is 15¼" high by 14½" wide, allowing for ¼" deep grooves in the hood sides. The sides themselves are made of oak, ½" thick, 14" high by 7¾" wide. The two frames shown in the next

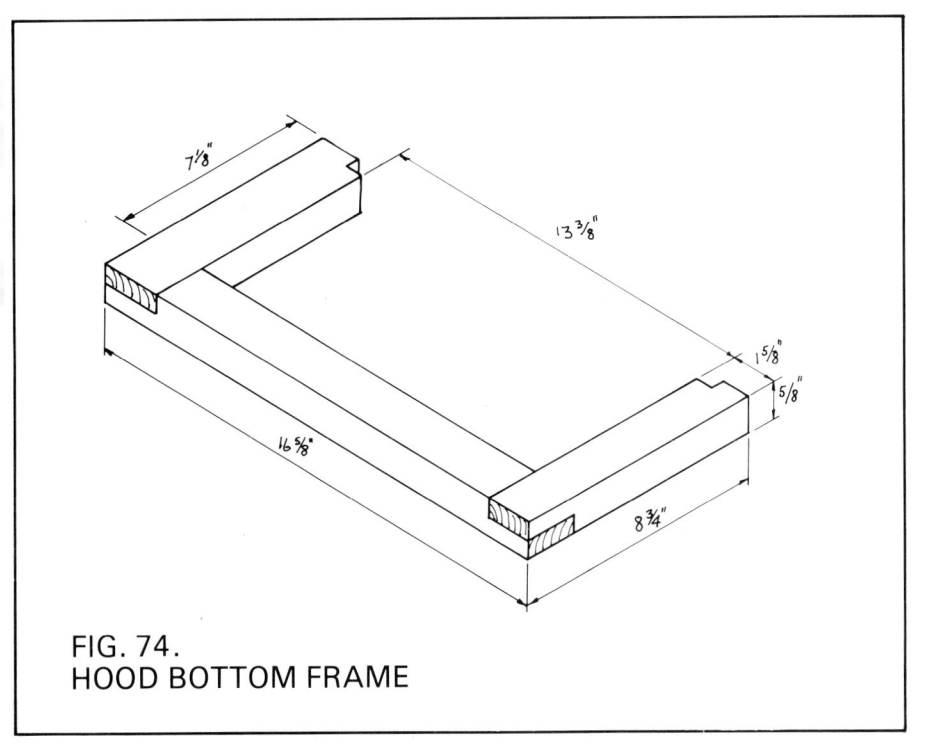

**FIG. 74. HOOD BOTTOM FRAME**

**FIG 75.
HOOD TOP FRAME**

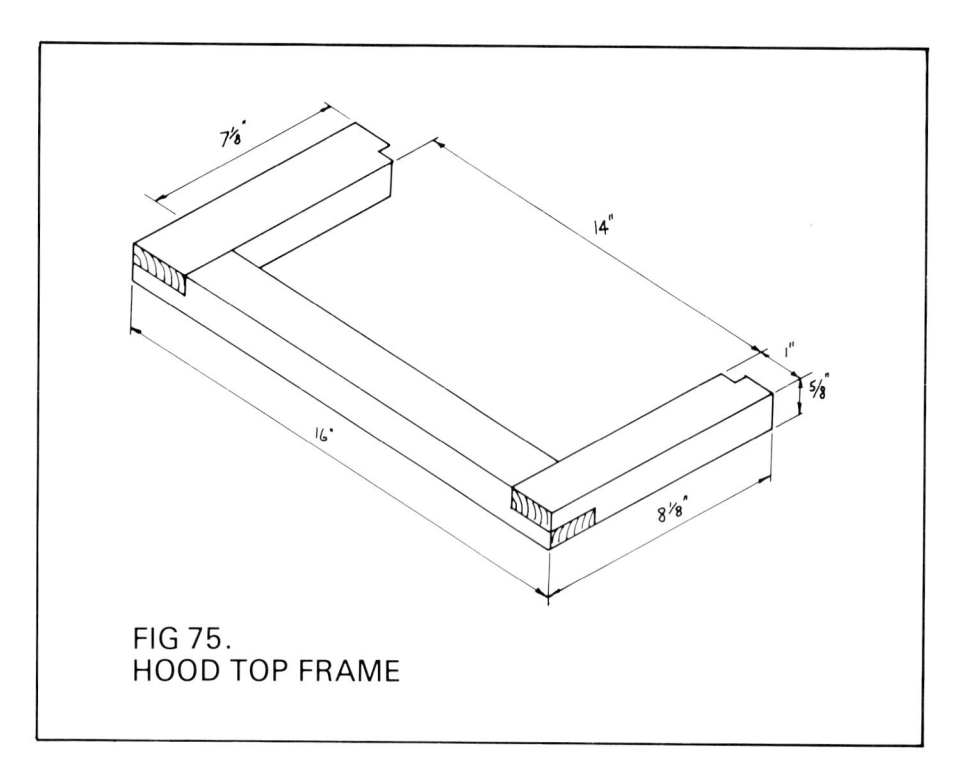

sketch of the hood assembly are planted underneath and on top of the hood sides. It is best to fit these using mortice and tenon joints into the frames. If you decide to do this, make the sides longer to allow for the tenons. Alternatively the sides may be dowelled to the frames. The grooves in the sides should be so positioned that the front face of the mask comes a full 5/8" in from the front edge of the sides to accommodate the door. The back edges of the sides are again rebated to fit over the backboard.

The next stage is to make up the two frames as in the drawing. These are made of oak, half lapped at the corners. The bottom frame will sit on the trunk top moulding, and will have a small moulding fitted round its outer edge to form the rebate which locates the hood on the trunk. Thus, the outer edge of the bottom frame should be about 1/16" proud of the trunk moulding all round to allow a little clearance for ease of assembly. As I mentioned earlier, this bottom frame projects slightly inside the hood sides to act as a runner, and the frame fits tightly back against the front face of the mask. To fit this frame, it is best to cut two or three tenons on the bottom edges of the side panels and mortices to match on the side legs of the frame. A perfectly sound alternative is to dowel the sides of the frame. Do not use a butt joint here as you will be glueing to end grain, and this will not produce a strong joint.

The general arrangement drawing of the case shows a window in each side of the hood. These windows serve two purposes. Firstly, they allow you to watch the line winding onto the barrels during winding without removing the hood — so you can see when the lines are fully wound. Secondly, I think it is a pity, after having done all that work on a movement, to then hide it in a wooden box! If you wish to make these windows, cut them before you glue the frame of the hood together. The bottom of each aperture should be just above the seatboard and the windows should be about 3" wide and 6" high. Cut the apertures and either rebate them on the inside to take the glass or glue and pin small quarter round oak moulding round the outer edge to form a rebate.

The top frame of the hood also fits back against the front face of the mask, but this time, the inner edges of the sides of the frame come flush with the inside faces of the side boards. Again, mortice and tenon or dowel joints should be used. When the whole assembly is glued up, tack a couple of bracing pieces across the back of the hood until it is finished, and stick three or four triangular glue blocks down the joints between the mask and the sides.

When the glue is dry, try the hood on the trunk, and trim the trunk sides to length so that the mask fits the dialplate perfectly and with the same overlap all round. You will then ensure that the finished hood will fit properly.

The remainder of the hood carcase is purely decorative, so the dimensions are not so critical, and the job is quite straightforward. The cross section drawing shows a frame of oak, 1/2" by 3 1/2", fitted round the outer edge of the top frame. This should have mitred corner joints so that there is no end grain showing when finished. This frame can be glued and screwed on from the outside as the decorative mouldings applied round its bottom edge will hide the screw heads. These mouldings can be built up from two parts as shown, and should again be accurately mitred at the corners. The top of the hood is a sheet of 1/4" ply, glued and pinned to the top edge of the hood. The edges of the top can be planed flush when the glue has set, and the top moulding is applied to hide the edges of the plywood. This then completes the basic construction.

Many oak cases had some form of carved decorative frieze round the top. This can be simulated by glueing small squares of oak about 1/4" thick round the hood just under the top moulding. Adjust the size of the blocks so that they space out evenly round the hood. Another nice touch which could be done is to fit a fretted panel across the top of the hood instead of the blocks of oak. Frets were often used for decoration and served the

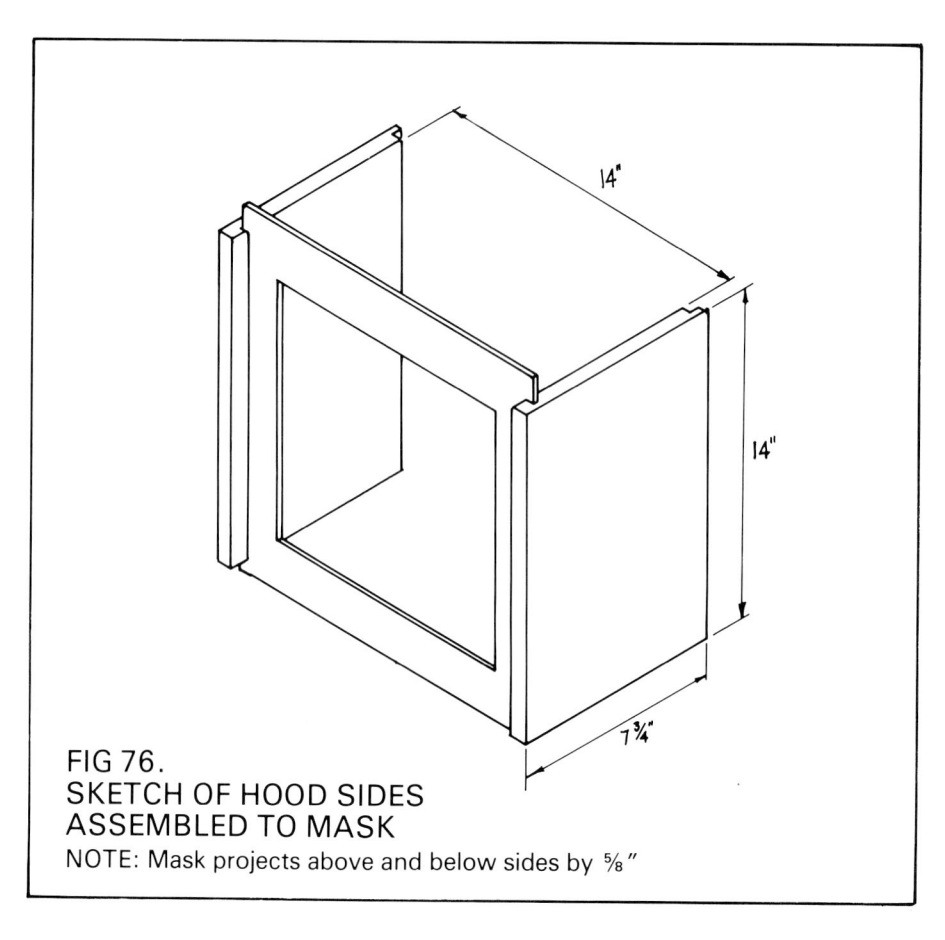

**FIG 76.
SKETCH OF HOOD SIDES
ASSEMBLED TO MASK**
NOTE: Mask projects above and below sides by 5/8"

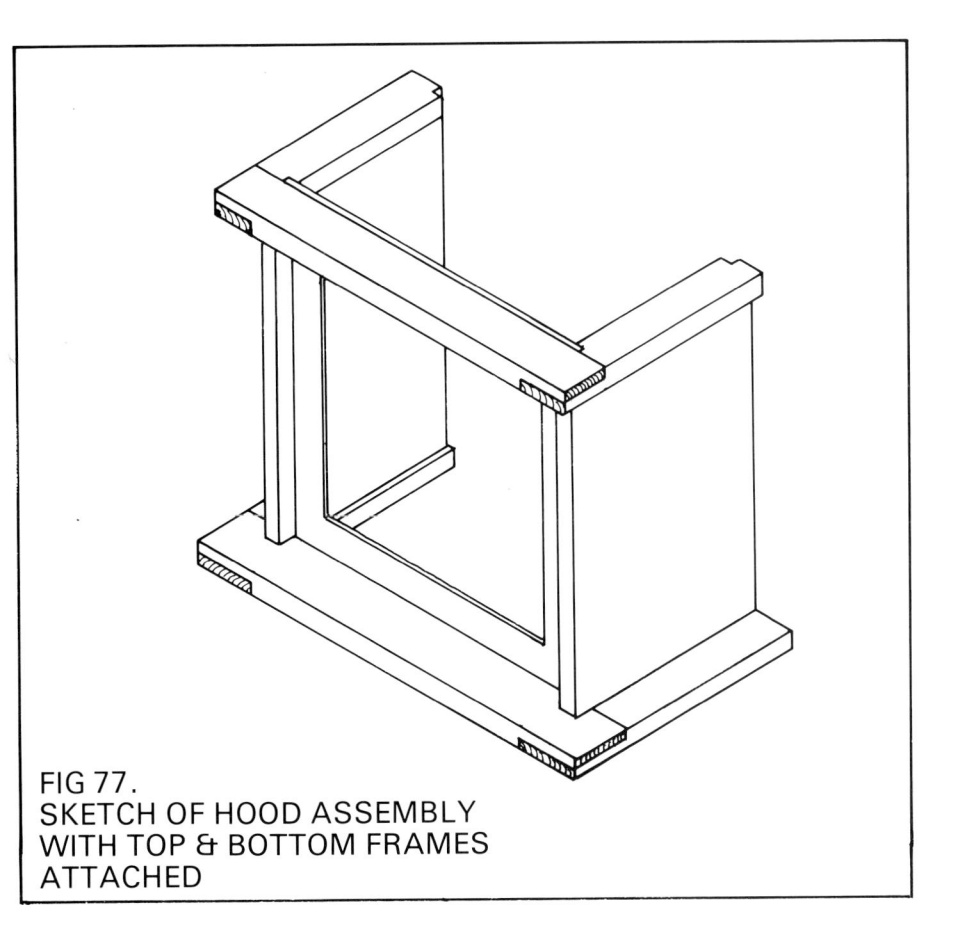

**FIG 77.
SKETCH OF HOOD ASSEMBLY
WITH TOP & BOTTOM FRAMES
ATTACHED**

ly figured oak which is well seasoned for the door. The door is one panel, $\frac{1}{2}''$ thick, $9\frac{1}{2}''$ wide and about 32" long. On the drawing of the front view of the case, I have shown some decorative shaping to the top of the door, as was common. All sorts of shapes were used here, so you don't necessarily have to make it the same as shown. After shaping, the front edge of the door is shaped with a scratch stock to form a decorative bead. Finally, the back edge of the door is rebated to fit the frame, leaving about $\frac{1}{16}''$ clearance all round, and the front face of the door about $\frac{1}{8}''$ thick. The trunk door is hung on brass 'acorn' hinges which are the same as ordinary brass butt hinges, but have decorative acorn shaped extensions on the hinge pin. These are the traditional design and are readily available. Finally, a small brass lock and escutcheon plate is fitted to the trunk door, and a small axe drop handle to the hood door.

### FINISHING

I am definitely no expert on wood finishing techniques, and I would advise you to read a good book on finishing methods. However, I can give some guidance.

The first decision to be made is whether you want to leave the case a natural oak

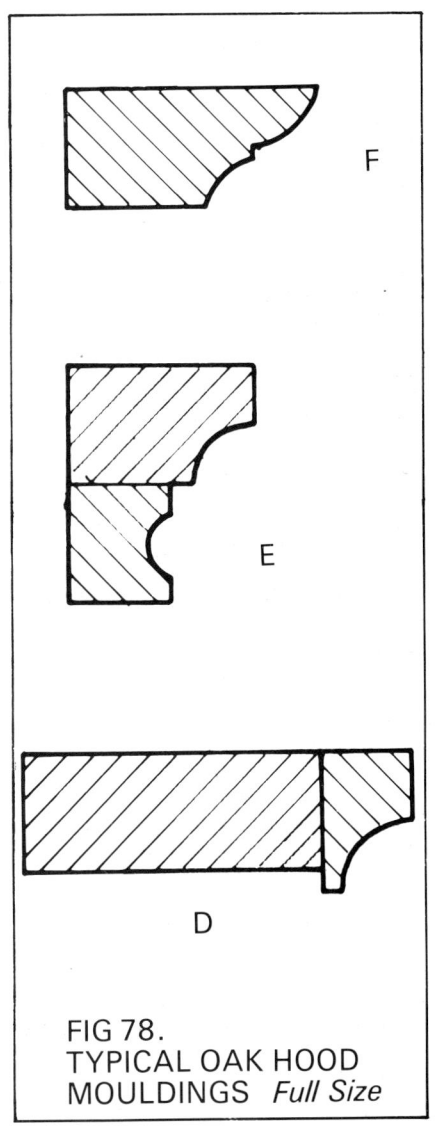

purpose of letting the sound of the bell out. If you decide to fit a fret, cut an aperture across the hood front, about 1" high and 14" wide. This should be rebated on the inside to take the fret which is held in place with pins. The fret itself is made by laminating three or four pieces of oak veneer, fretting out a design which was usually floral, and glueing red silk over the back of the fret both to keep the dust out and to present a pleasing appearance.

The hood door is made of $\frac{5}{8}''$ thick oak, with either half lap, mitred bridle, or mortice and tenon joints at the corners. The door's outside dimensions should be just a little smaller than the square frame created by the hood sides and the frames, top and bottom. The aperture for the glass should be the same as the hole in the mask. A rebate is formed round the inner edge of the frame for the glass. As with the side windows, this can be planed out with a rebate plane before the frame is glued up, or it can be formed by glueing and pinning oak beading around the aperture after the frame has been assembled. The latter method is probably the easiest.

This door is hinged with special swan neck hinges made from $\frac{1}{16}''$ brass sheet. These allow the door to pivot outwards clear of the pillar, and are very easy to fit. When you make the hinges, remember to countersink the screw holes on opposite sides to make a pair! The hinges are let into the top and bottom edges of the door as shown in the photograph and are screwed into place with countersunk screws. The door is then placed in the frame and small brass pins are tapped through the pivot holes of the hinges into the hood frames to act as pivots. Hanging the door is best left until after the hood has been polished.

The columns and brass capitals which are fitted either side of the hood door are also best left until the polishing has been completed. The capitals are available from material suppliers, and you will need four of these with a $1\frac{1}{4}''$ square base. These are fitted onto the ends of turned oak columns which are tapered from about 1" diameter at the bottom to about $\frac{3}{4}''$ at the top. These columns can be left plain or fluted as you wish. The columns need to be about 14" long to allow for trimming and fitting into the capitals. Ready turned columns are also available, but if you decide to turn your own, do not turn a straight taper. They should 'bulge' slightly and have a small curve along their length, otherwise they will appear to taper to a waist at their middles, and this effect looks most odd. The Greeks discovered this when they built their temples with rows of columns. The solution they found in making the columns bulge slightly is called entasis. (It is remarkable what you pick up making clocks!) The finished columns and capitals are glued together with an epoxy resin adhesive, and are fixed to the hood with four brass pins through holes drilled in each corner of the capitals. It is best to lacquer all the brasswork to prevent tarnishing as it is difficult to polish this on the case.

When the hood is finished, lay the trunk on its back on the flat bench and fit the hood. Through the dial aperture, draw a line round the outer edge of the rebate to mark out the final shape of the backboard. Cut the board to size and re-fit the hood to make sure you have got a good fit. If all is well, remove the backboard from the trunk, apply glue to the mating faces and screw or pin the back permanently into place.

The final part of the case is the trunk door. As I mentioned when discussing timber selection, you need a piece of nice-

**FIG 78.
TYPICAL OAK HOOD
MOULDINGS** *Full Size*

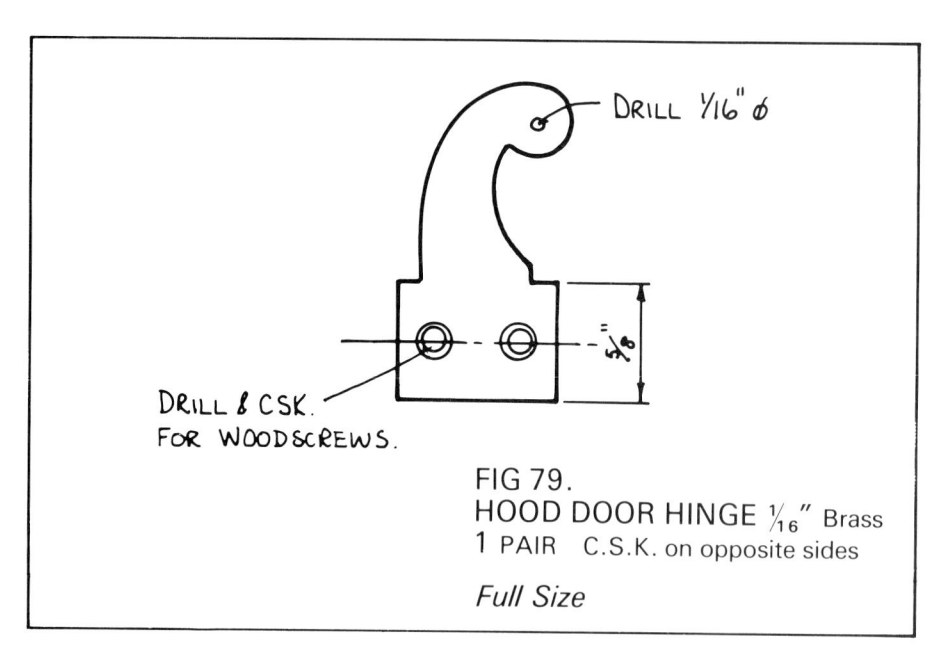

**FIG 79.**
**HOOD DOOR HINGE 1/16" Brass**
**1 PAIR** C.S.K. on opposite sides
*Full Size*

DRILL 1/16" ⌀
DRILL & CSK. FOR WOODSCREWS.

colour, or whether you want to darken it with stain. The vast majority of antique cases are dark, but I cannot say whether this was so originally, or whether the cases have darkened with age. Some years ago, I had a case made and polished without staining, and this case is now a nice golden honey colour — and gets better as time goes by.

If you wish to stain the case, there is a wide selection of stains on the market — 'Colron Wood Dye' being a popular one. These dyes are available in a wide range of colours, so test them out on a piece of scrap timber to see the effect. As we do not want a 'piano finish', do not use a grain filler before staining — simply brush the stain on evenly, or apply it with a piece of cloth. Allow a a day for the stain to dry thoroughly, and then rub down all over with fine glasspaper until you have achieved a really smooth surface. As with all finishing methods, the final finish can only be as good as the preparation.

After sanding, wipe all traces of dust off the job and from now on, if possible, work in a warm, dust free atmosphere. A good method of removing dust is to use a 'Tak Rag' which is available from car body finish suppliers. These are slightly sticky and pick up the last trace of dust. Don't forget to check that all pin heads have been filled and rubbed down.

There are many different polishes available, including French Polishing kits, You will require the normal transparent polish. The 'Furniglas' kits are good, and full instructions are supplied. If you buy French Polish on its own, brush on a couple of coats all over the case as thinly as possible and leave it to dry. Then rub down using very fine wire wool and repeat the operation. Continue doing this — applying the polish and rubbing down thoroughly each time — until you have built up a good, smooth coat of polish. Very lightly, rub this down again with fine wire wool, before completing the job with a 'rubber' (a cloth pad about the size of an egg). This 'rubber' is made up from fine cloth (old, clean mutton cloth is good), with cotton wool or similar wrapped up in it to form a pad with a smooth face. Charge the pad with french polish, thinned with a little methylated spirit, and gently rub all over the job — firstly with a circular motion, followed with a figure of eight movement. Keep the pad damped with polish and go over the work until a shine builds up. The experts finish off with the rubber dampened with spirit alone, after letting the polish dry for a day. This is a very delicate process and requires a good deal of skill — so practice on some scrap before you attempt it.

When you have finished applying the polish, let it dry off in a warm atmosphere for a few days, and then it will be ready for wax polishing. There is only one polish to use, and that is beeswax. This can be bought ready made up, but it is best if home made. This is done by gently melting a cake of beeswax in a tin placed in boiling water and then stirring in pure turpentine. You will have to experiment with the mixture to obtain the consistency of thick cream when cold.

With this wax, there is no substitute for hard work or 'elbow grease'! Apply the wax with a soft cloth a little at a time and then burnish the surface with a clean, soft duster. You will find that, with time, this waxing will produce a beautiful rich surface. This finish improves with further wax polishing over a period, creating that deep lustre so admired in good quality antique furniture.

The final jobs are the fitting of the glass to the hood, and the brasswork to the case. The glass is fitted in the usual way with putty on the inside of the frame. It is best to darken the putty to match the wood and make it less obtrusive. The brasswork should be refitted taking care to use a good screwdriver so that you don't slip off the screws. Escutcheons, etc., are usually fitted with small brass pins.

The movement is attached to the seatboard with two hooks bent up from 1/8" diameter steel rod. These hook over the bottom pillars, pass through holes drilled in the seatboard, and are then pulled tight with 1/2" square brass nuts threaded 5 BA,

*Plate 62. Detail of hood hinge and capital.*

screwed onto the projecting ends of the hooks. Before drilling the holes for the hooks, position the movement on the seatboard, and adjust its position so that the dial is central in the mask and so that the mask touches all round the edge of the dialplate. Gently remove the hood taking care not to disturb the movement, and mark the position of the holes for the hooks. It is also a good idea to mark the position of the bottom corners of the frontplate on the seatboard so that the movement can be replaced accurately in the case. Don't forget to drill two holes for the ends of the lines — these holes should be so positioned that the weights hang equally spaced centrally in the trunk.

Fit the movement and lines to the seatboard, position the case against a wall and check that it is vertical — side to side and front to back. As the case will almost certainly stand against a skirting board, cut a piece of wood for packing to go between the back of the case and the wall. It is best to plug the wall and put a large screw through the backboard and packing, into the wall, especially if the clock is standing on a carpet. If you want to stand the clock in a corner, fit a diagonal bridging piece across the corner and screw the clock to this.

Finally, hang the weights and pendulum, wind up and check if the pendulum is in beat. If not, bend the crutch as described previously until the tick is even. Set to the correct time, check the striking and there you are!

**THE MAHOGANY CASE**

As I mentioned at the beginning of the section on casemaking, the mahogany case is not a project for the inexperienced. However, anyone with a reasonable knowledge of cabinetmaking should be able to build this case from the information given on the drawings and in the text for the oak case.

The basic carcase construction is identical to the methods used for the oak case, but the break arched case is more complicated because of the curved mouldings, etc. These are best produced on a wood turning lathe as a full circle on a faceplate. This will give you two sets of mouldings, so you may be able to share a set with a friend. The case should be made from the best quality mahogany available, but for the best visual effect, the trunk door and base panel should be veneered with mahogany curl — preferably applied with hot glue and a veneer hammer.

The curved top panels of the pagoda top hood can be made up from strips of wood glued to the hood frame and sanded to shape before veneering, or they can be thin ply, glued and pinned to the frame and then veneered. Details of the mouldings are given, but alternatives are quite acceptable.

I hope that readers have or will enjoy making this clock, which will provide many years of pleasure, and should undoubtedly become a 'family heirloom'. I also hope that I have taken some of the mystery out of clockmaking and that I have encouraged a few people to take up a most interesting and absorbing hobby. As I warned in the beginning, clockmaking is definitely infectious!

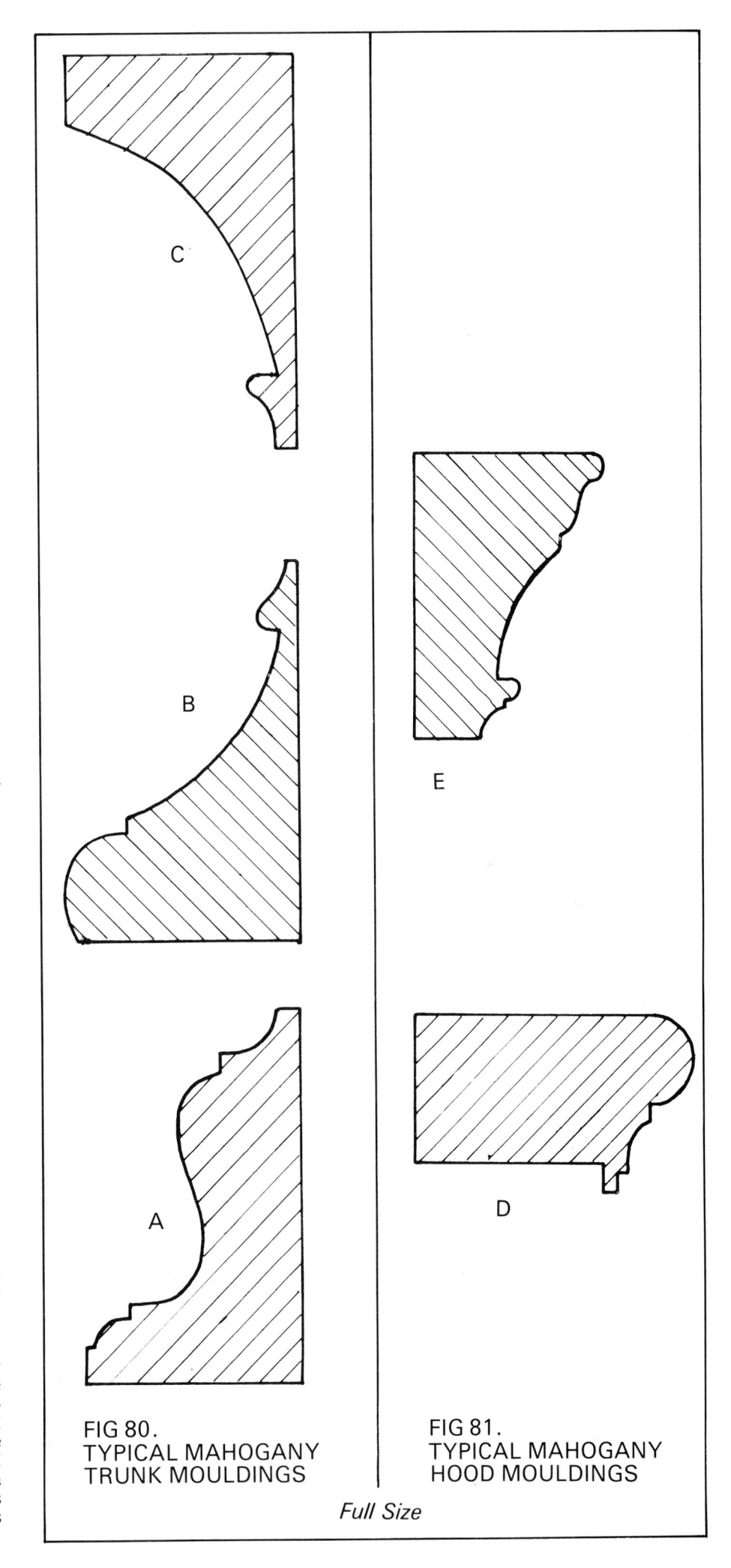

FIG 80.
TYPICAL MAHOGANY TRUNK MOULDINGS

FIG 81.
TYPICAL MAHOGANY HOOD MOULDINGS

*Full Size*

FIG 82.
THE MAHOGANY BREAK ARCH CASE

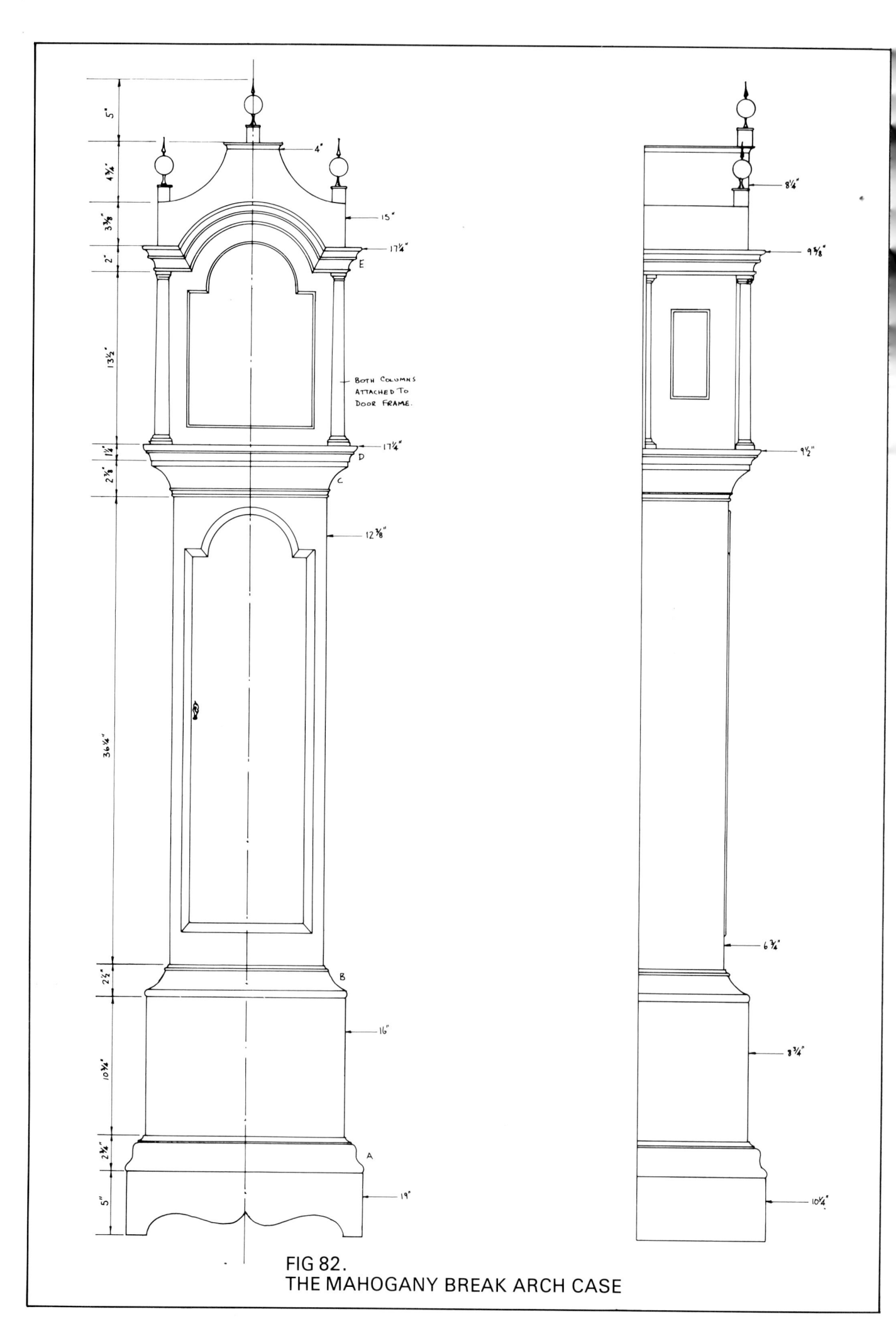

# Appendix I

## GEARING THEORY & METHODS OF CUTTING CLOCK GEARS

At first sight gearcutting seems, to the beginner, to be a mysterious and complicated art. There appear to be so many problems and calculations involved that many people are put off altogether before they have considered all the alternative methods open to them. The other factor which has put a good many people off clockmaking is the lack of reasonably priced equipment available commercially for those who prefer to make models, clocks etc., rather than spend their time making tools to make models with, and never getting round to building the finished article. Happily, the last problem is a thing of the past. It seems that every month now, new items of much sought after equipment are appearing on the market, to cater for the rapidly expanding clockmaking and restoration market. I hope in the following pages to dispel some of the problems faced by the average model engineer, by explaining the various ways of cutting clock wheels and pinions, describing a few pieces of equipment that can be made in the workshop, methods of making simple cutters and bringing to your notice equipment and accessories available from suppliers. I cannot hope to cover every alternative, but I hope to provide enough information to enable anyone with the average range of tools and machinery to put together a reasonable set up for making clock gears. Of course, model engineers are renowned for their ingenuity and I expect many will be able to dream up all sorts of devious ways round some of the problems!

### The Theoretical Aspects Of Clock Gears

When considering tooth forms used in clockmaking, two factors stand out that make the Clockmaker's task much easier. The first is that, as clock gear trains always revolve in one direction only, we are not too bothered by backlash. Secondly, except in one instance, the wheels always drive the pinions. Together, these two facts simplify the problem as these conditions allow us to use a very simple tooth form. Another factor is that generally, the amateur is only making one clock, so the wheels and arbors are 'planted' between the plates using a depthing tool to mark out the pivot hole positions, after each pair of arbors has been adjusted for mesh in the tool. Thus, unlike batch production, one off situations allow a little latitude on tolerances within reason.

The most common tooth form used in clockmaking is called Epicycloidal. This has been developed over the years in an effort to reduce friction and to establish a standard for the Horological Industry. Observation of a number of antique clocks will soon reveal a very wide range of tooth shapes and proportions. Indeed, it is unusual to find identical teeth in two clocks by different makers. Some antiques have the crudest tooth forms but have worked perfectly well for 200 years or more. A good form will, however, reduce friction and thus the driving power required, which will in the long term reduce wear on pivots, etc. When you consider the methods available to the Clockmaker in the 1700's, with crude hand operated machinery, you will realise how much easier it must be nowadays.

Nowadays, the pitch of clock gears is measured by the metric MODULE system. The PITCH of a wheel is the distance between the tip of one tooth and the tip of the next. The PITCH CIRCLE DIAMETER is the effective diameter of a wheel or pinion, ie., the line along which the teeth mesh. The Module refers to the pitch diameter of a wheel in millimetres divided by the number of teeth. Thus, a wheel with a pitch diameter of 30 mm and 60 teeth will be 0.5 Module. The larger the module, the bigger the teeth. This is the reverse of the Imperial Diametral Pitch system, which refers to the number of teeth on a wheel with a pitch circle diameter (P.C.D.) of 1". A wheel of 34 D.P. and 34 teeth will have a P.C.D. of 1". A wheel of 34 D.P. with 68 teeth will have a P.C.D. of 2" dia. etc. Therefore, the higher the D.P., the smaller the teeth.

The P.C.D. can be considered as the diameter of a plain roller. If we want a ratio of 2:1, we need a pair of rollers, one say 1" dia. and the other 2" dia. If we press these two rollers together and revolve the larger one, the small roller will revolve twice for each rev of the large one. Plain rollers, however are liable to slip, so we need to have teeth projecting from them so that they mesh together. The amount the teeth project beyond the Pitch Circle is called the ADDENDUM. Thus, the diameter of the blank needed for a wheel must be bigger than the P.C.D. by an amount equal to twice the addendum.

As cutters are supplied in Module form nowadays, I shall use this system from now on. However, I should point out that as my machines are calibrated in good old 'thous', I will always convert from metric to imperial. I will also mention here that a pocket calculator is invaluable when working out gearing and conversion calculations. Although the formulae are straightforward, some of the numbers involved get a bit awkward at times! If you are used to the D.P. system, conversion is easy:-

$$\text{MODULE} = \frac{25.4}{\text{DIAMETRAL PITCH}}$$

Vice Versa

$$\text{D.P.} = \frac{25.4}{\text{MODULE}}$$

Fig I shows the tooth form for Epicycloidal wheel teeth. All the dimensions shown are given as constant factors of the Module. It is thus possible to work out all the dimensions once the Module to be used has been decided. From the diagram you will see that the tooth width, the space width and the radius

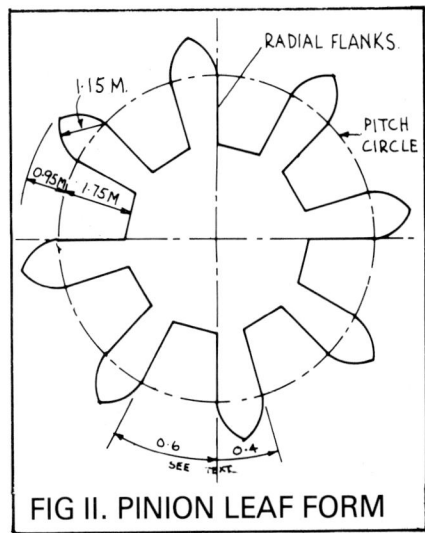

**FIG I. WHEEL TOOTH FORM**
M = MODULE — see text.

**FIG II. PINION LEAF FORM**

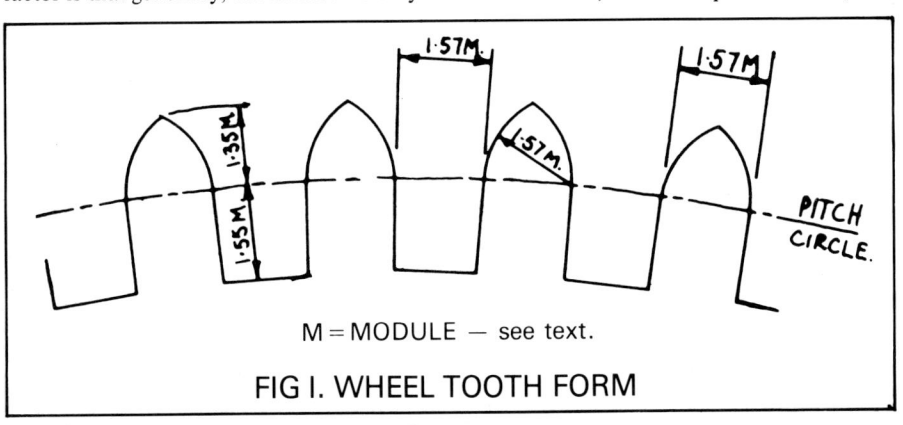

of the tip curvature are all the same ie. 1.57 x Module. The addendum is given as 1.35 x Module and the Dedendum 1.55 M. This information, as I will explain later is necessary if you wish to make your own cutters. As the numbers given are constants for all modules, the addendum will always be 2.7. Thus the formula for the outside diameter of any wheel blank is:-

O.D. = (NUMBER OF TEETH + 2.7) × MODULE.

As an example, for a wheel of 0.8M, with 60 teeth:-

60 + 2.7 = 62.7.
62.7 × 0.8 = 50.16 mm.
$\frac{50.16}{25.4}$ = 1.975".

The Dedendum is not critical as long as it is no less than that given. The teeth can be deeper and often are. The only point to bear in mind is that excessively deep teeth are weaker at the root. It is worth mentioning here that some modern cutters produce a tooth form with a semi-circular root, in line with modern engineering practice, as this leaves no sharp corners from which cracks could start. Antique clocks always have square roots to the teeth.

Fig II shows the form of pinion teeth or 'leaves'. Unlike wheel teeth, there are many theoretical variations according to the application for which the pinion is to be used. The form shown here is a good general purpose form for the average work the amateur is likely to come across. In theory, wheels (ie. gears with more than 15 teeth) should have teeth with radial flanks. In practice, this can be ignored with no detrimental effect. Pinions, however, are a different matter! As you can see in the diagram, pinion teeth taper quite steeply, ie. they have radial flanks, and the angles vary considerably on pinions with different numbers of teeth because they are of small diameter. This explains why pinion cutters are more difficult to make and why different cutters are needed for different numbers of leaves.

The addendum, dedendum and tip curvature are still factors of the Module. However, if you buy commercial cutters, check with the supplier on the addendum, as this does vary. The other major difference between wheels and pinions is that the spaces are wider than the leaves. This is done to provide the necessary clearance when a wheel and pinion mesh. The ratio of leaf thickness to space width is 0.4:0.6. ie., the distance between the centre of one leaf and the centre of the next may be divided into 0.4 for the leaf and 0.6 for the space. From this information, obtained from Britten's Watch and Clockmaker's Handbook, Dictionary and Guide (a most informative book which I recommend to anybody with an interest in Clockmaking), we can calculate the angles of the flanks of pinion leaves for any number of leaves.

As an example, I will work out the angles for an 8 leaf pinion as illustrated. First, divide 360 by 8, which equals 45 degrees. This angle must now be split into the ratio of 0.4:0.6, so divide it by 10 which gives us 4.5. The angle occupied by the leaf is therefore 4.5 x 4 which is 18 degrees, and the space is 4.5 x 6 or 27 degrees. The cutter for an 8 leaf pinion must have a tip that makes an included angle of 27 degrees. For the form shown, the formula for calculating the outside diameter of the blank is:-

OUTSIDE DIAMETER = (NUMBER OF LEAVES + 1.9) x MODULE.

If you buy commercial multi-tooth cutters, the above information, apart from the addendum, is only of academic interest as the tooth form is already decided for you. As I will explain later, if you are able to work out the dimensions of the teeth

*Plate I. Commercial multi-tooth cutters. No. 1 Escape wheel cutter, No. 2 0.75m, 7 leaf pinion cutter. No. 3 0.75m, 8 leaf pinion cutter. No. 4 0.8m, wheel cutter. No. 5 o.4m, Swiss H.S.S. wheel cutter.*

and leaves it is possible to make your own cutters and thus save expense.

## PRACTICAL METHODS OF WHEELCUTTING

There are many ways of producing these theoretical tooth forms on a blank in the workshop. Some methods are better or quicker than others, but the model engineer with the average workshop should be able to provide a set up to suit his purposes without too much expense. As I said earlier, there is an ever growing range of equipment and accessories coming onto the market. more of which later. It is also possible to make most of the bits and pieces needed for yourself. In my workshop I have a Myford lathe, a small Centec horizontal mill, an Emco Mentor Mill, a bench drilling machine and the usual small grinder, buff etc. Until about two years ago, I did all my wheelcutting on the lathe, using home made accessories. This set up produced perfectly good wheels and pinions but was sometimes inconvenient, so I bought a second hand Centec. Fitted with a home made dividing head, this machine now cuts all the wheels and pinions I make. I hardly ever take the dividing head off it, as I rarely use it for anything else. The Mentor Mill is ideal for gearcutting, as is the Unimat lathe. As demand has increased, I have recently bought a Swiss Wyssbrod pinion cutting machine which is completely automatic. This machine is obviously outside the range of the model engineer's workshop. It's hardly worth setting it up to cut less than 10 pinions, so for one offs, I still use the methods available to the average modeller.

To cut gears, we must provide three things, no matter which method we decide to use. These are:—
1. A suitable cutter to produce the tooth form required.
2. A holder for the cutter, mounted in a spindle driven by a motor.
3. A method of holding the blank and rotating it in precise steps. (Dividing or indexing.) I will deal with these in order.

## 1. CUTTERS

The best cutters, especially for pinions, are undoubtedly the commercially available multi-tooth type. (see Plate 1.) These are available from suppliers in a wide range of sizes, in either carbon steel or

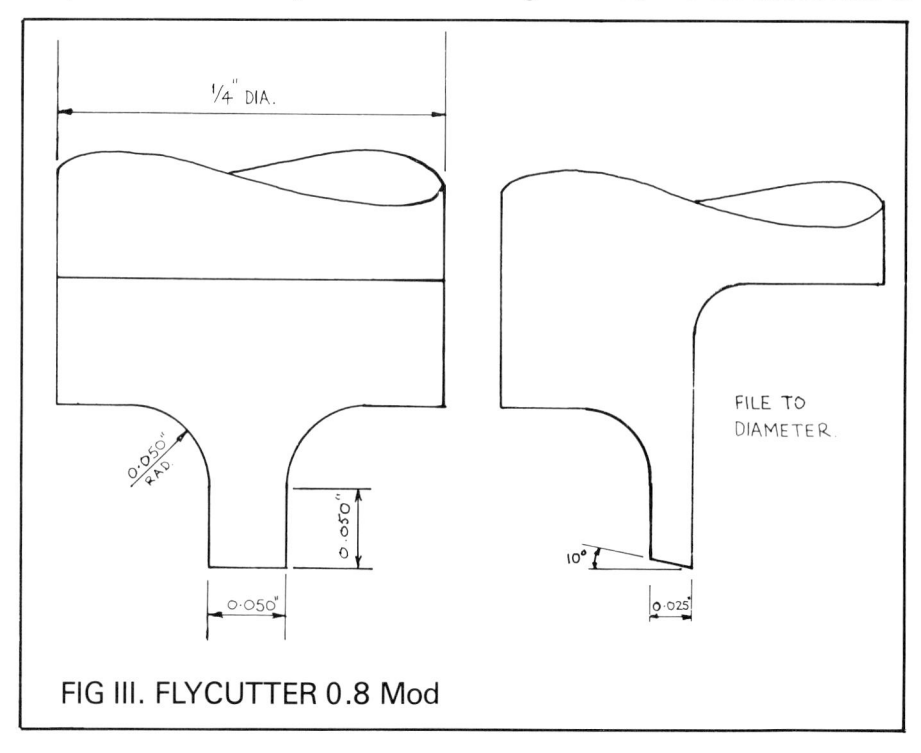

FIG III. FLYCUTTER 0.8 Mod

H.S.S., and are well worth buying if you intend to get involved in Clockmaking. (I warn you now, Clockmaking is infectious!) Unlike Involute cutters, Epicycloidal cutters will cut a wide range of tooth numbers. For all practical purposes, one wheel cutter will cut all numbers usually found in clock trains. Pinions, however, are a different matter. As we saw earlier, the angle of the leaf flanks varies with the number of leaves, so we need a different cutter for each number of leaves. The cutters are, of course, supplied in Module sizes. The manufacturers will supply data on addenda etc., and they recommend that pinions are cut with a cutter 0.05M smaller than the wheels, but on a blank with the O.D. calculated using the same Module as the wheels. This produces a thicker root to the pinion leaves and makes the pinions to a form which is stronger and more in keeping with antiques. Brass wheels can be cut very effectively with home made silver steel fly cutters, which are very easy (and cheap) to make. Plate II shows a group of three fly cutters for cutting wheels, escape wheels and ratchets. I use fly cutters a lot, especially for making replacement wheels for old clocks with odd tooth forms, when the old wheel can be used as a template to make a cutter. Referring again to Fig I, all the information needed to make a cutter is there. As an example, I will describe how to make a 0.8M cutter. Fig III shows the form of the tip.

The three critical dimensions are all the same, ie. 1.57M. The addendum will come out right if the tip radius is correct, and the dedendum is not critical and can easily be 1.57M as well, so all the dimensions are the same. To find this measurement, simply multiply 1.57 by 0.8. This comes out to 1.256 mm., which can be converted to 'real inches' by dividing by 25.4. The answer is 0.0494" so 0.050" will be near enough. We now need a lathe tool with its left hand cutting edge ground to a radius of 50 thou. Few of us have tool and cutter grinders, so a simple gauge can be made by drilling a No.39 hole in a piece of scrap. Saw and file to the hole leaving a quarter of a circle which will be your gauge. Grind a piece of ¼" square H.S.S. to shape and hone the tip with a stone until it fits the gauge perfectly.

The cutter is made from ¼" dia. silver steel. Hold this in the chuck or a collet and set the tool exactly to centre height. Face the end of the rod and then turn a spigot 50 thou dia., with a parallel length 50 thou long. The shape of the tool will produce a shoulder of exactly the right radius. The next stage is to mill or file the formed end of the rod to exactly half its thickness, as in the drawing. This is one of those jobs where the roller filing rest is very useful. Check with a micrometer and file down until the tip is just over 0.025" thick leaving a little to stone off later. It is sometimes necessary to reduce the diameter of the cutter from the rod, leaving it about 1" long. Hardening and tempering is done as usual. Heat the tip to red heat and quench in cold water or preferably a 10% brine solution. Polish the flat cutting face on a fine oilstone and very gently, heat the shank of the cutter away from the tip. Carefully watch the polished face until it turns to a light straw colour. As soon as this happens, quench again. Finally, using a fine oilstone, give the tip about 10 degrees clearance angle and stone the face to a high polish with nice sharp edges. Once you get used to making these cutters, the whole job takes about ten minutes. I never throw them away and I now have a box of about 50 or so and can usually find one just right for the job.

Cutters of this type will cut many wheels if they are used properly and treated carefully. If they are revolving fast enough, they will cut to a full depth in one pass leaving a nice finish with very little burr on compo brass. A good speed for fly cutting is about 4,000 R.P.M. and upwards. The faster the better. The most likely time for the cutter to break is on first contact with the blank, so take care and feed slowly.

Unfortunately, simple fly cutters as just described will not cut steel pinions successfully, their tips are just too weak to stand the strain. It is possible to make fly cutters for pinions but they are slightly more complicated to produce. The best method I have seen is described in a very informative booklet published by Chronos Designs Ltd., called 'Wheelcutting, Practical Notes For Clockmakers.' (See list of suppliers.) I am indebted to Mr. Isaacs of Chronos for allowing me to describe this method).

Make a blank from Gauge Plate (carbon steel), 1½" to 2" dia. and 3/16" to ¼" thick. Drill a hole in the centre of the blank and mount it on a mandrel in the lathe. Machine the circumference to the required form for either wheels or pinions, calculated as already described. (See Fig IV). Saw four sections from the disc as shown by the dotted lines, clean up with a file and drill a mounting hole in each cutter so that it can be mounted on an arbor like a disc type cutter. File on some top rake like a lathe tool and heat treat just the same as the simple fly cutter. This method produces four identical cutters with much stronger tips, which are in fact identical to one tooth of a multi tooth cutter. Four off may seem excessive but there are three major advantages with this system. Firstly, the cutter tip is far less likely to break and you have identical spares if it does. There is nothing more frustrating than cutting half of a 96 tooth wheel and then breaking the cutter! Besides having to stop to make a new one, the wheel could easily be ruined. Secondly, they can be re-sharpened without altering the form, by simply stoning the front face. Finally, if you have a friend who is also making clocks, each of you can make different cutters and swap two of each.

Of course, to cut steel pinions the spindle speed must be dropped considerably to about 150 to 200 R.P.M., using a slow feed and several passes, with plenty of coolant. Some people prefer to gash the pinion blank to full depth with a small slitting saw first, followed by a form cutter to finish the leaf form. This method certainly reduces the strain on the cutter. As pinions should be hardened, most people use silver steel for the blank. This material is notoriously difficult to machine, but free cutting silver steel is available and this reduces the problem considerably. (See list of suppliers). I have used mild steel and case hardened it with success. Another alternative if you have any 'contacts' is cyanide hardening which is very good. The Chronos booklet also shows methods of making multi-tooth form relieved cutters which I haven't tried so I can't comment on them, but they certainly look worthy of trying.

## 2. MOUNTING AND DRIVING CUTTERS.

Having either obtained or made the cutters, we need to provide a means of driving them. Traditionally, Clockmakers

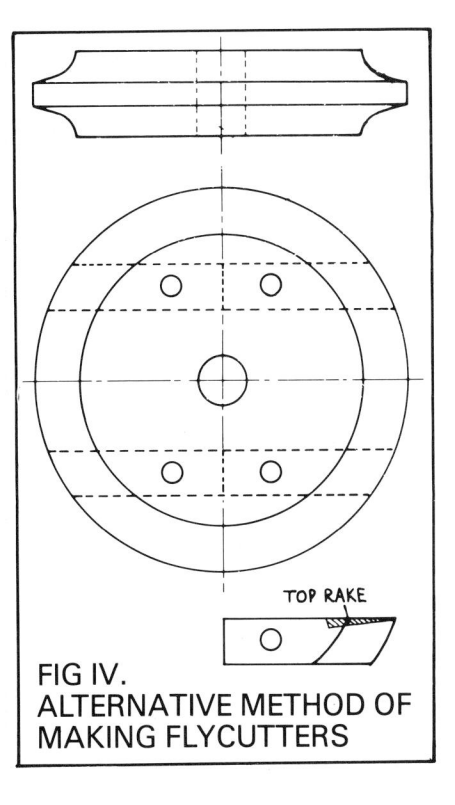

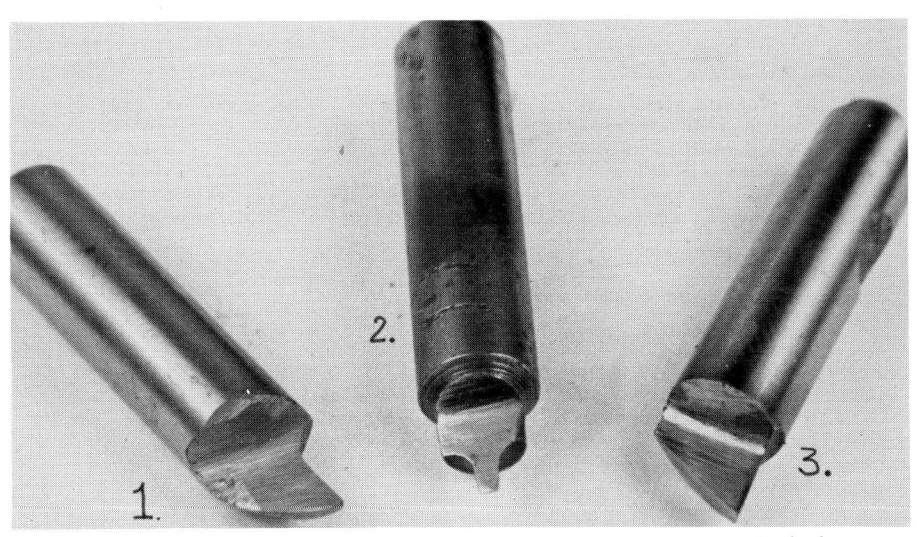

*Plate II. A group of single point flycutters. No. 1 Escape wheel cutter. No. 2 0.8 mod. wheel cutter. No. 3 Ratchet cutter.*

**FIG IV. ALTERNATIVE METHOD OF MAKING FLYCUTTERS**

used a wheelcutting engine, examples of which are now Museum pieces and are collectable in their own right! I was recently loaned an engine, manufactured by Chronos, which I demonstrated at the Leicester Exhibition. (See Plate III). The most striking impression on using this machine for the first time was the ease of setting up and using it and the speed with which it produces wheels. I seem to remember spending a good deal of time producing blanks which it digested all too quickly! As you can see, the set up is very simple and demonstrates the requirements for virtually all wheelcutting set ups. A spindle, revolving in a sleeve, carries an accurate division plate at one end, and can be fitted with a range of work holding devices at the other end. Thus, the blank and division plate are locked to a common spindle and can revolve together. A spring detent arm is fixed to the base of the machine to pick up the holes in the plate.

The cutter, either a fly cutter or multi tooth type, is held in a spindle which revolves between adjustable centres in a cutter frame, and is driven by a small motor via plastic belt. There is provision for centering or offsetting the cutter, adjusting the depth of cut and feeding the cutter into the job. Chronos supply cutter frames and spindles as separate items (see Plate IV), and I believe they also supply ball race spindles.

Most people, however, will not be able to justify buying a purpose built machine just to cut wheels and pinions (unless they get the bug!), so how can normal engineering machinery be adapted to suit our purpose? If you have a milling machine, this is the obvious answer. My millers have No.2 Morse taper spindles and I have a Clare collet chuck to hold endmills. This chuck has a ½" dia. collet, so I make my cutter arbors with a ½" shank, with various ends to suit the cutters I use. For fly cutters, I use ¾" dia, with a ¼" hole at the opposite end. An axial hole, drilled and tapped 2 B.A. takes an Allen screw which clamps the cutter in position. For circular cutters, I use ½" dia. rod and turn the end down to suit the bore of the cutter. A 2 B.A. nut and washer hold the cutter in place. I have fitted my Centec with pulleys which will give a maximum speed of 2,000 R.P.M. which is a little slow for fly cutters but works. This is an ideal speed for commercial cutters used on brass. Using a milling machine, the blank is held on a dividing head on the machine table, which I will describe later.

If you haven't got a mill, don't despair. As I said earlier, I cut wheels on the lathe for several years. The most convenient set up is to hold the blank on a mandrel in a collet, on the lathe headstock. A dividing head or index plate is fitted to the back end of the headstock spindle using an expanding plug similar to Fig V. A detent fixed to any convenient part of the lathe completes the indexing set up, which in effect is exactly the same as the wheel-cutting engine turned on its side. For normal clock wheels, the cutter can be held in a frame or spindle held in the four tool turret, with the cutter revolving in a horizontal plane set exactly to centre height. With this method, the depth of cut is controlled by the cross-slide and the cutter is fed in by traversing the saddle. The spindle is driven by a small motor of about ⅒th H.P., fixed to a baseboard and clamped to the rear of the lathe bed. Drive is

*Plate III. A Chronos wheel cutting engine.*

transmitted by stretchy plastic which can be heat joined to any length. The stretch easily accommodates spindle movement, thus avoiding complicated motor mounts.

Holding the spindle in the tool post is not the best method, as there is no provision for offsetting or centering the cutter easily. The best method is to mount the spindle on a vertical slide fixed to the cross/slide, so that the cutter passes over the top of the blank. This is a far more versatile arrangement. The cutter is still driven as just described, but the depth of cut is controlled by the vertical slide and centering or offsetting is done using the cross-slide. Cowell Engineering Ltd. have recently produced some excellent attachments for their range of lathes. Plate V shows their milling and wheelcutting attachment with its self contained motor unit, mounted as I have just described on their vertical slide. The unit is very neat, can be mounted on any of its four faces and there is provision for two speeds, 300 or 1800 R.P.M.

I tried this set up at the Leicester Exhibition and found that it certainly does the job very efficiently. They had it set up with a Chronos index plate and detent on the back of the headstock and a sensitive lever feed arrangement to the saddle mounted in the leadscrew housing. Blanks were held on a mandrel, and pinions could easily be held between centres. The lathe bed has a tapped hole placed centrally between the shears into which a support can be screwed to steady slender arbors whilst cutting. Cowells hope to produce more much needed attachments in the near future so it is worth watching their range of products.

Arrand Engineering also produce a range of milling and drilling spindles which are ideal for this application, or you can make your own. There must be as many designs for spindles as there are model engineers! Much has been published in the past and many suppliers stock kits, castings, etc., so a look through the adverts should help.

*Plate IV. A Chronos cutter frame.*

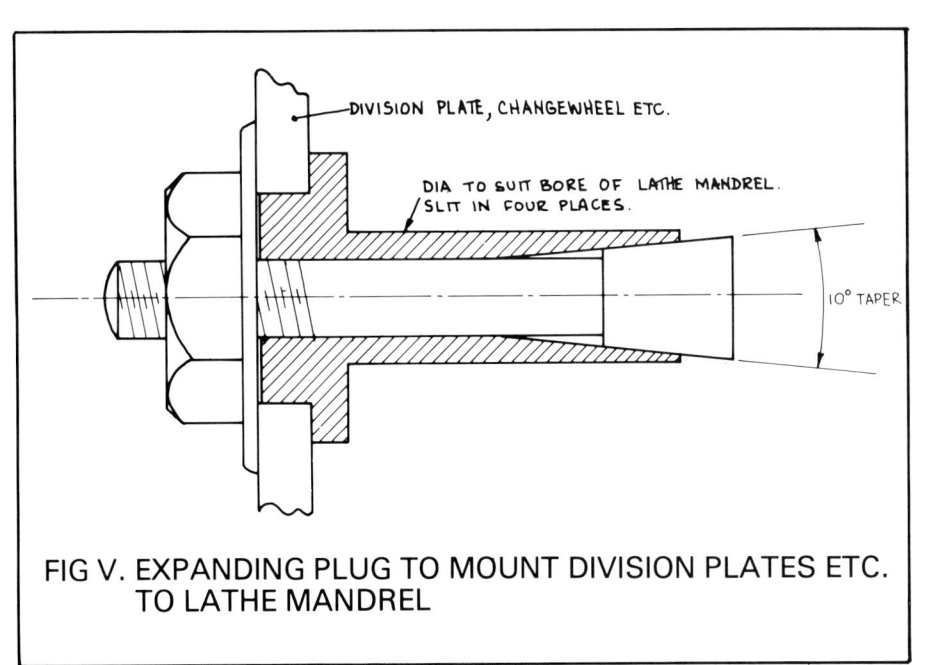

**FIG V. EXPANDING PLUG TO MOUNT DIVISION PLATES ETC. TO LATHE MANDREL**

## INDEXING.

This is another subject which has been discussed many times in the past so there is a lot of information available. There are designs for dividing heads to fit direct to the bull wheel of Super 7 lathes, attachments to fit to the rear of lathe mandrels and no end of other gadgets which can be adapted to suit our purposes.

The most versatile method is the dividing head with 40 or 60:1 reduction. With a suitable range of plates, a vast range of divisions are possible. Dividing heads are expensive and often too big, so I designed and made my own. The spindle nose is identical to the Myford, so I can use all my Myford collets, etc. This is a great time saver as I can swap blanks from the lathe to the miller easily and accurately. The body of the head is fabricated from mild steel sections. The reduction unit is a 60 tooth Myford changewheel, which meshes perfectly with a 6 T.P.I. Acme thread, to give 60:1 reduction. (See Plates VI & VII). I can also fit a 40 tooth wheel, or an index plate to do direct indexing. If there is the demand, I will prepare an article on this head and try to organise castings etc.

I have four plates for the dividing head, all 4½" dia x ⅛" thick mild steel with a 1" dia centre hole. Plate 1 has 39, 33, 21, 20, 18 & 15 holes, Plate 2 has 27, 23, 19 17 & 16, and plate 3, 49, 31 and 14 plus space for any extras I may need sometime. All the plates are home made and with care, accurate plates can be made using changewheels as masters. The method I use is as follows:-

Make the blanks and mount them on a mandrel turned 'in situ'. Using the expanding plug shown in Fig V, mount the required changewheel on the back of the lathe mandrel and arrange a detent fixed to the lathe. I use a steel bar bolted to the lathe, carrying a bolt with a pointed end which fits between the changewheel teeth. The holes can be drilled direct on the lathe using a drilling spindle on the cross slide. A B.S.3 centre drill is ideal as it is rigid, centre punching is not needed and if you arrange a stop on the bed, the holes can be countersunk at the same time. The other way, if you haven't got a drilling spindle, is to make a holder to fit the tool post with a ¼" hole drilled and reamed in position in the lathe so it is exactly at centre height, like a boring bar holder. Make a centre punch from a bit of ¼" dia. silver steel and slide this into the holder. Adjust the cross-slide so that the tip of the punch is at the correct radius. You can now accurately centre punch the blank as you index it round. Only a light tap is needed. When you have marked all the holes, punch them deeper on the bench and carefully drill on the drilling machine using a B.S.3 drill.

If you want to use direct indexing, you will obviously want your plates to have the same number of holes (or multiples) as the numbers of teeth you wish to cut. Again, Chronos make a wide range of plates to suit all needs at a very reasonable price and they will make special plates to order. Cowells also make a simple indexing device and a neat little dividing head.

After such a lengthy description of gadgets, you will be pleased to hear that once you have decided on a method which suits you and you have collected the gadgets you need, wheelcutting is very easy.

With the blank mounted on a suitable mandrel, and the spindle and indexing mounted in position, proceed as follows. Centre the cutter exactly (I use a ½" dia. bar with a 60 degree point held in a collet), and adjust the depth of cut so that it is too shallow. Switch on and take a cut. Index round to the next tooth and cut again. This should produce a tooth with a flat top. Feed the cutter down in small steps, indexing back and forth between these two first cuts until the tip of the tooth is fully formed with the slightest 'witness' left on the tip. An

*Plate V. The Cowell wheel cutting and milling attachment.*

*Plate VI. Home made fabricated 60:1 dividing head.*

eyeglass is useful here. When you have the depth correct, lock all the slides not being used and proceed round the blank cutting each tooth to full depth at one pass. It is surprising how easy it really is and if you prepare all the blanks you need first, a complete set of wheels can be made very quickly.

I hope that this description has answered a few questions and what's more important, I hope I haven't put anybody off the idea of making a clock. Individual set ups are quite straightforward, but as you will realise, trying to cover all the eventualities in one article is not so easy!

*Plate VII. Home made dividing head showing 60 tooth change wheel meshing with Acme thread. Thread can be disengaged to allow free rotation of spindle for direct indexing.*

# Appendix II

## List of Suppliers and Useful Addresses
(In alphabetical order)

**ARRAND ENGINEERING LTD**
The Forge, Knossington, Oakham, Leics.

Milling Spindles, mandrels, etc.

**THE BRITISH HOROLOGICAL INSTITUTE**
Upton Hall, Upton, Newark, Notts.

**CHRONOS DESIGNS LTD**
The Old Barn, Woods Lane, Potterspury, Towcester, Northants.

Wheel cutting engines; pinion cutting machines, cutters, cutter spindles and frames, division plates, wheel blanks, kits of brass, piercing saw tables. (Catalogue).

**COWELL ENGINEERING**
Oak Street, Norwich.

Cowell lathes, drills, milling machines and accessoris.

**GOODACRE ENGRAVING LTD**
Lodge House, Wyvern Industrial Estate, Long Eaton, Notts.

All kinds of engraved clock dials, name Cartouche, date rings, hands, spandrels, etc. (Catalogue).

**G. K. HADFIELD**
Blackbrook Hill House, Tickow Lane, Shepshed, Loughborough, Leics.

Just about everything for antique and new clocks, also a very good stock of books. Case mouldings and brasswork. (Catalogue).

**SYDNEY G. JONES**
8 Balham Hill, London SW12 9EA

All types of good quality files, including pivot files.

**MYFORD MACHINE TOOLS LTD**
Beeston, Nottingham.

A comprehensive range of lathes and accessories suitable for the clockmaker. (Leaflets).

**RICHARDS OF BURTON**
Woodhouse Clockworks, Swadlincote Road, Woodville, Burton on Trent.

Clock dials, hands, spandrels, wheel and pinion cutters, comprehensive range of clock parts. (Catalogue).

**NATHAN SHESTOPAL LTD**
1 Grangeway, London NW6 2BW.

A large range of tools and accessories and clock parts. Agents for Emco lathes and Chronos. (Catalogue).

**RITA SHENTON**
148 Percy Road, Twickenham, TW2 7JG

A large range of horological books. (Lists).

**J. SMITHS & SONS LTD**
42-54 St Johns Square, Clerkenwell, London EC1P 1ER

All types of brass and non ferrous metals.
Also branches throughout the country (DELTA METALS).

**ALAN TIMMINS CLOCKS**
6 Trent Crescent, Attenborough, Beeston, Notts.

Kits of ready cut wheels and pinions.

**P. P. THORNTON (Successors) LTD**
The Old Bakehouse, Upper Tysoe, Warwickshire CV35 0TR

Wheel, pinion, ratchet and 'scape wheel cutters. (Lists).

**J. M. WILD**
12 Norton Green Close, Sheffield, S8 8BP

Clock depthing tools, mainspring winders, spandrels and castings. (Catalogue).

**COLIN WALTON CLOCKS**
11 Tythe Close, Gazely, Suffolk.

Depthing tool kits.

**LEONARD BALL**
44 Market Street, Lutterworth, Leics.

Clock parts, case mouldings, case fittings.

*NOTE:* Catalogues or lists are available where stated. Some of these are free (send an SAE), whilst there is a charge for others. Very often, the cost of the catalogue is deducted from the first order.

# NEW from EME
# It's the size of lathe you've been waiting for!

# The Compact 5

Larger than a Unimat smaller than the Compact 8. Height of centres 65mm, distance between centres 350mm.

Send to London office for FREE literature with full specifications:

**EME LTD.**
B.E.C. House,
Victoria Road,
London NW10 6NY
Tel: 01-965 4050

**EME LTD.**
18 Sidcup Road,
Roundthorne Industrial Estate,
Wythenshawe,
Manchester 23
Tel: 061 945 2382

*Please send me details of the Compact 5*

Name_____
Address_____
_____
_____

EME / BEC
THE ELLIOTT GROUP
T.C.B.

---

## VANCO No. 1 LINISHER
### KIT OF PARTS £95.00
**COMPLETE AND READY TO ASSEMBLE**

★ 1" to ¼" wide abrasive belts all ex stock 14 grits.
★ 8½" × 6" Cast Iron Ground Angle Table.
★ V Belt Drive.
★ Grind, Smooth, Polish, Alloy, Brass, Steel, and HS Steel, Wood with perfect control.
★ Complete with Assembly Instructions. Less Motor.

*Manufactured in the U.K.*
*Price includes p&p*

**ACROVU LIMITED**
6 Burners Lane, Kiln Farm Ind. Est., Milton Keynes MK11 3HB
Tel: (0908) 563465

---

## Goodacre Engraving Limited

(Clock Parts Section)

Lodge House, Wyvern Industrial Estate,

Long Eaton, Nottingham

*Telephone Long Eaton (06076) 4387*

We are specialist hand engravers of all types of dials, engraving by hand in the traditional manner to restore dials to their former glory or produce excellent reproduction dials to the customers specifications — as seen illustrated in this book.

We also have a Clock Parts Section now firmly established, having gained a reputation for stocking exceptionally high quality BRITISH MADE clock parts and castings at very competitive prices.

*A photographic catalogue and price list is available — 'phone or write to the above address for details.*

# Cowells 90 CW Lathe
## ALL–BRITISH PRECISION CLOCKMAKERS LATHE

Justly renowned for its precision and fine traditional craftsmanship, this lathe probably represents the finest value for money in its field today.

- *Infinitely variable speed control up to 4000rpm through thyristor controller*
- *Accepts 8mm collets in both headstock and tailstock*
- *Lever operated drilling attachment*
- *Cross slide and compound slide with toolpost*
- *Continuously rated motor*

Write or phone for our literature which we will be pleased to send to you

**Cowell Engineering Limited**
Oak Street Norwich NR3 3BP England
Phone (0603) 614521 Telex 975205

Tooling available for the Cowells 90CW Lathe includes:-

- *Full range of collets – 0.5mm to 6.5mm*
- *Adaptors to accept 3 and 4 jaw chucks, faceplate etc.*
- *Indexing unit with range of dividing plates*
- *Saddle lever feed attachment*
- *Motorized milling spindle for gear and pinion cutting*
- *Range of cutter arbors for gear cutting*
- *Tipover tool rest*
- *Precision roller filing rest*

United States Sales:
**Cowells Incorporated**
P.O.Box 427,
226 East Adele Court,
Villa Park, Illinois, 60181.
Phone (312) 279–0490.
Telex 72–1586.

---

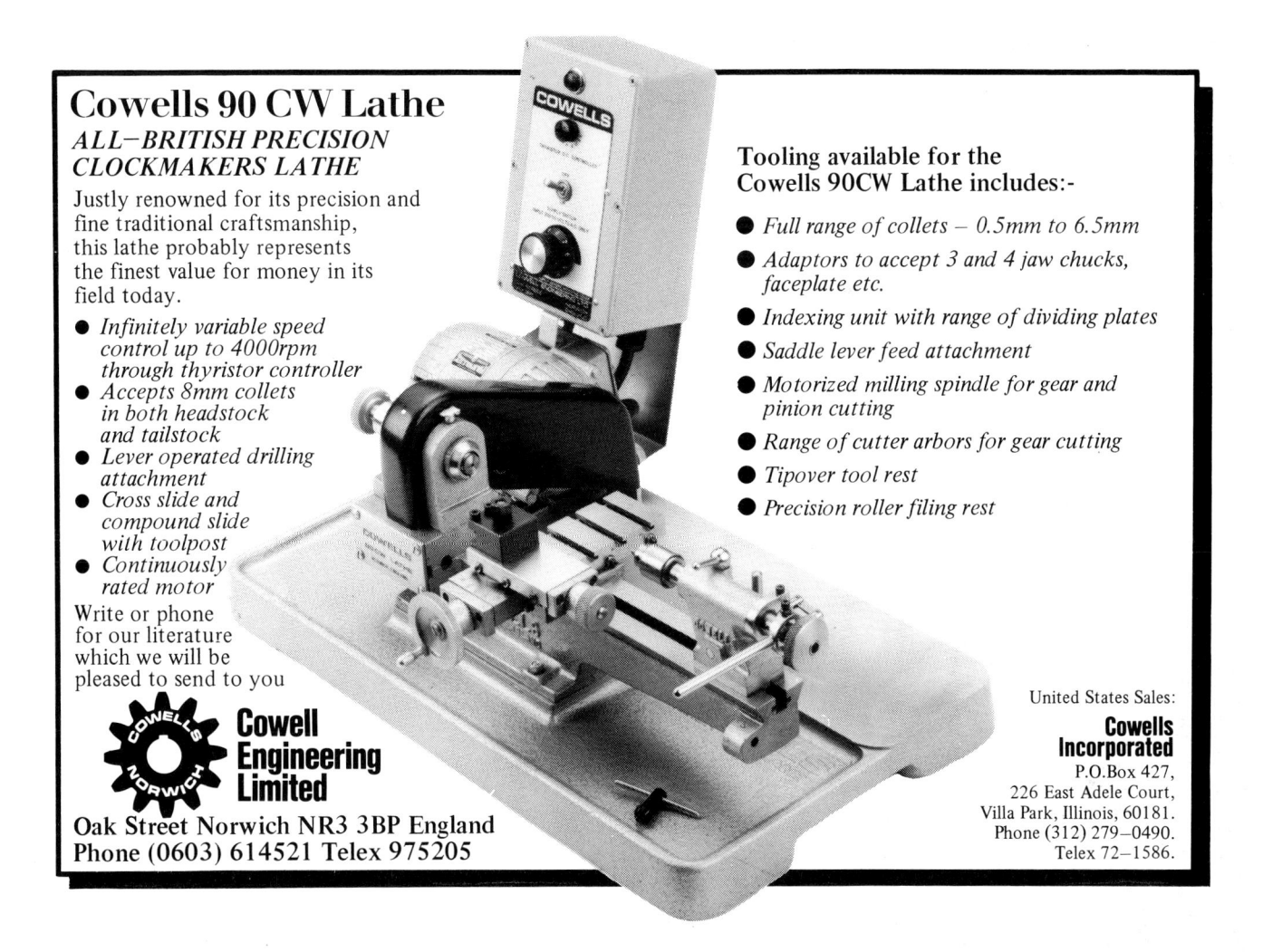

# CRAFTS FOR 4 SEASONS
1120 Melton Road, Syston, Leics. Tel: 607242
On the A607 to Melton from Leicester.

Complete supply of materials for the clockmaker
Brass Discs any size cut in ¹⁄₁₆″ up to ¼″.
Brass Plates in ⅛″ standard sizes.
Brass Tube for barrels

Complete packs cut to size.
Our own design of case for wall clock to take
8 day long pendulum movement.
All machined for easy assembly in
Oak and Deal.
Painted white dial faces.
Brass embellishments.

*Send 30p for detailed catalogue.*
**Closed all day Mondays.**

---

## *DO YOU HAVE A DIAL AND MOVEMENT THAT NEEDS A CASE?*

**WE CAN MAKE YOU A GOOD CLOCK CASE TO THE SAME PROPORTIONS AND STYLE AS THE OLD ORIGINALS, IN GOOD SOLID WOOD OR WITH BEAUTIFUL VENEERS.**

Don't leave that dial and movement hidden away, get it recased now and have the pleasure from it. We make clock cases to suit all dial sizes, in Oak, Mahogany, and Walnut. All hand made by craftsmen, on traditional lines. Dials repainted, and old movements and cases completely restored. Established dealers in Antique Longcase Clocks. Send for details and colour photographs enclosing £1.00

**The John Dann Clock Company**
20 High Street, Branston
Lincoln, England.
Tel: 0522-791710

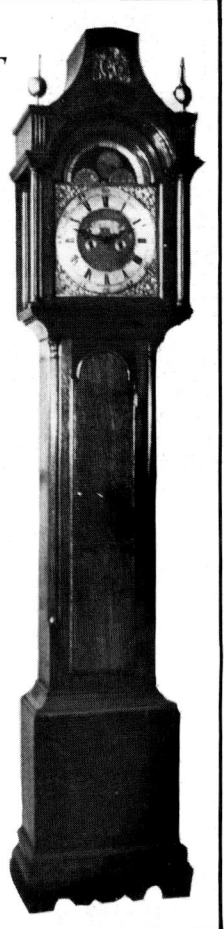

# Established 1907
# Shesto

## For the finest range of Gear Cutters and Hobs!

We are stockists of all gear cutters featured in the John Wilding "How to Make" series of books and we stock a very wide range of standard and traditional square bottom tooth cycloidal cutters. Send for our NEW FREE leaflet explaining how to calculate and order the correct cutters to suit your requirements.

Your cheque/P.O. made payable to: NATHAN SHESTOPAL LTD.
Send to: Dept. **T.C.B.** Nathan Shestopal Ltd. 1 The Grangeway, Kilburn LONDON NW6 2BW Tel: 01-328 3128

```
Please send me your FREE leaflet on gear
cutters.                              Please tick ☐
Please send me your catalogue of
Watchmakers and Modelmakers tools.
I enclose £1.00.                      Please tick ☐
Name
Address
                                              T.C.B.
```

## DELIVERY EX-STOCK
# SIMAT 101
### 2" x 12"
### CENTRE LATHE

## "MORE LATHE FOR YOUR MONEY"

S.A.E. FOR DETAILS

**Alphabeta Engineering**
GAYMERS WAY INDUSTRIAL ESTATE,
NORTH WALSHAM, NORFOLK.
Telephone 06924 3302.

## THE BOYLE WHEEL and PINION CUTTING ENGINE
*Manufactured by* **Boyle (Engineering)**
13 Beechwood Avenue, Frome, Somerset, England    Tel: 0373-66059

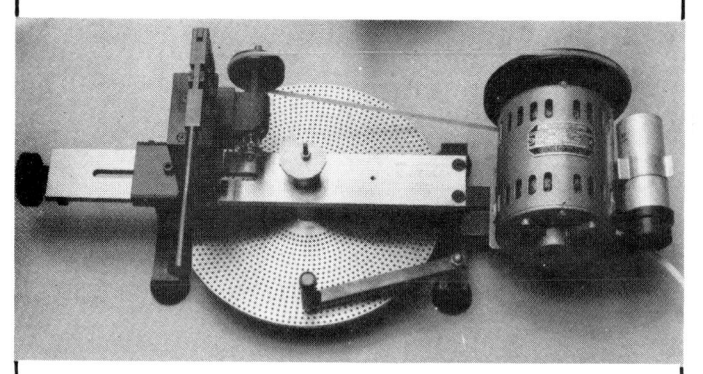

The Boyle Wheel and Pinion Cutting Engine is a precision made machine and will enable the amateur or professional to cut wheels and pinions on the one machine. In addition it is possible to cut change wheels for a small lathe. The following operations can easily be carried out on the machine:—
★ cutting clock wheels and pinions
★ cutting change wheels for small lathes, ⅜" thick, from 1" up to and including 6" diameter
★ making a count wheel or snail
★ milling squares on round stock
★ milling key way 1¼" long

**A DRILLING ATTACHMENT IS AVAILABLE, AS AN ADDITIONAL ACCESSORY**
and will allow you to carry out the following operations:—
★ direct indexed drilling for clock lantern pinions
★ direct indexed drilling for pin wheel
★ any other indexed drilling within the 69 divisions
★ indexing exact divisions for seconds and minutes on small dials.
*The price of this additional accessory obtainable on request.*

## SKELETON AND GRANDFATHER CLOCK KITS FOR ALL SKILLS

Wide range of models and kit types, many requiring no horological knowledge. All available with gears and pinions ready cut for those who require it.

**Model Range**
Passing Strike Skeleton
Hour Strike Skeleton and Long Case
¼ Chime West/Whit. Skeleton and Long Case — 8 Bells
Congreve Clock

**Kit Types Requiring:**
either (a) Hand tools and polishing only, or
(b) Lathe work, drilling and polishing, or
(c) Gear-cutting, lathe work, drilling and polishing

*Please specify model(s) and kit type for free literature or send £2.00 for our comprehensive Catalogue.*

**CLASSIC CLOCKS**
Dept. E/1 (Clerkenwell)
Jerusalem Passage
St Johns Square
London EC1V 4JP
Tel: (01) 2514027

Hour Strike Skeleton Clock

Congreve Clock built from fully machined kit

---

## Traditional Clockmakers Depthing Tools

**High quality brass depthing tools complete with cone and lantern runner in fitted wooden box.**

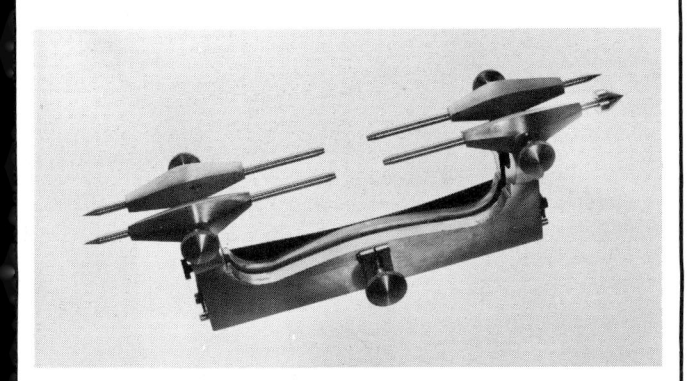

Send for catalogue showing lantern clock parts, spandrels, depthing tools, mainspring winders, clock testing stands, and books.

### J M W (CLOCKS)
12 Norton Green Close, Sheffield S8 8BP.
Tel: 0742-745693

---

## CHRONOS

The wheel cutting tool specialists can supply all the wheel blanks and many other materials for all the Alan Timmins Designs.

*Write for our catalogue describing our range of wheel cutting engines, pinion mills and division plates, etc.*

### CHRONOS DESIGNS LTD
Woods Lane, Potterspury,
Towcester, Northants
Tel: (0908) 542395

*Rack strike, long case, clock movement.*
*Clockmaker: Alan Timmins of Nottingham.*

# Built with a Myford

Talk lathes to Model Engineers and clockmakers and you'll be talking Myford. It's been that way for many years.... Why? Simply because they're professional machines, built by professionals and used by professionals. And unlike many lathes currently available, a Myford will extend your skills, a full range of equipment is available to enable it to perform almost any task.

A precision job requires a precision machine, ask around, look around and then, like so many craftsmen before you, come to Myford.

Details of the complete Myford range can be obtained by writing to Myford Ltd., Beeston, Nottingham NG9 1ER.
Tel: Nottingham (0602) 254222.

**Myford** Machine Tools

**the choice of the craftsman**